Lecture Notes in Mathematics

A collection of informal reports and seminars
Edited by A. Dold, Heidelberg and B. Eckmann, Zürich

86

Category Theory, Homology Theory and their Applications I

Proceedings of the Conference held at the Seattle Research Center of the Battelle Memorial Institute, June 24 – July 19, 1968
Volume One

1969

Springer-Verlag Berlin · Heidelberg · New York

Library of Congress Catalog Card Number 75-75931 Printed in Germany. Title No. 3692

Preface

A conference was held at the Seattle Research Center of the Battelle Memorial nstitute during the summer of 1968. The object of this conference was to bring to-ether research workers in the fields of category theory and homology theory and those ho applied the results of these theories to their own mathematical disciplines within lgebra or topology. Thus this was not, and was not intended to be, a tightly special-zed conference in categorical algebra (by comparison with the Midwest Seminars), the xpectation of its organizers being that the roles of category theory and homology heory within mathematics would emerge the more clearly from the conference and that he interplays of these theories with other parts of mathematics would be highlighted.

There were about 80 participants in the conference; at least that number of people, ully deserving of invitations, had unfortunately to be turned away owing to consider-tions of available space and optimal size for a working colloquium of four week's uration. Three invated courses of lectures were delivered:

J. F. Adams Generalized Cohomology

D. A. Buchsbaum Regular Local Rings

F. W. Lawvere Hyperdoctrines

n addition, about 40 seminar talks were given by conference participants. All those iving talks were invited to submit manuscripts with a view to publishing the pro-eedings of the conference. The number of manuscripts received (together with their ength and somewhat random arrival time) suggested the desirability of publishing the roceedings in several volumes, of which this is the first. Following the table of ontents of this volume we have, however, appended a list of titles of papers to appear n subsequent volumes; this list may, naturally, require amendment if further manuscripts ecome available after the publication of this volume.

I would like to take this opportunity to express, on behalf of the organizing ommittee, our deep gratitude to the Battelle Memorial Institute for their support of he conference and to the administrative and clerical staff of the Institute for their nvaluable assistance, in innumerable respects, in the running of the conference and the preparation of manuscripts. We were, indeed, fortunate to be invited to hold our conference under the auspices of Battelle. That the conference participants fully shared the sense of gratitude of the organizing committee towards the Battelle Memorial Institute is set in evidence by the following resolution, passed at the closing session of the conference, which I was charged to make public through the medium of the conference proceedings.

"This conference at the Battelle Memorial Institute, with its initial object Categories, Homology and their applications, has now reached its termination. Though our subjects are far from complete, the exchange of ideas about them at this conference has been intensive, coherent, lucid, and never petty, while the local facilitie provided by the Institute have been infinitely productive. As participants in this conference, we would like to thank the Battelle Memorial Institute and its most helpful and efficient staff for the extaordinarily fine facilities and services which have been placed at our disposal. We are equally indebted to the enthusiasm of the organizing committee and the wisdom of the scientific staff at Battelle for their provision of this ideal setting for the natural transformation of ideas into theorems."

Cornell University, Ithaca, November, 1968 Peter Hilton

Table of Contents

Papers to appear in future volumes

J. F. Adams	Generalized cohomology
J. Beck	On H-spaces and infinite loop-spaces
H. B. Brinkmann	Relations for groups and for exact categories
S. U. Chase	Galois objects
P. Dedecker	Three-dimensional non-abelian cohomology for groups
R. R. Douglas, P. J. Hilton and F. Sigrist	H-spaces
D. B. Epstein and M. Kneser	Functors between categories of vector spaces
D. B. Epstein	Natural vector bundles
P. J. Freyd	New concepts in category theory
J. W. Gray	Categorical fibrations and 2-categories
K. W. Gruenberg	Category of group extensions
R. Hoobler	Non-abelian sheaf cohomology
M. A. Knus	Algebras graded by a group
M. Karoubi	Foncteurs dérivés et K-théorie
F. W. Lawvere	Hyperdoctrines
F. E. J. Linton	Relative functorial semantics
S. MacLane	Foundations for categories and sets
E. G. Manes	Minimal subalgebras for dynamical triples
P. May	Categories of spectra and infinite loop spaces
P. Olum	Homology of squares and factorization of diagrams
J. E. Roos	Locally Noetherian categories
F. Ulmer	Kan extensions, cotriples and André cohomology

COALGEBRAS IN A CATEGORY OF ALGEBRAS

by

Michael Barr

Let $\underline{A}$ be a category and $\mathbb{T} = (T,\eta,\mu)$ be a triple on $\underline{A}$. Then we may form $\underline{B} = \underline{A}^{\mathbb{T}}$, the category of $\mathbb{T}$ algebras. There is then an adjoint pair $\underline{A} \underset{U}{\overset{F}{\rightleftarrows}} \underline{B}$ and we may ask whether or not F is cotripleable. More explicitly, we may form a cotriple $\mathbb{G} = (FU,\varepsilon,F\eta U)$ where $\varepsilon\colon FU \longrightarrow 1$ and $\eta\colon 1 \longrightarrow UF = T$ are the adjointness morphisms. Then there is a natural functor $\Psi\colon \underline{A} \longrightarrow \underline{B}_{\mathbb{G}} = \underline{C}$ and we are asking whether Ψ is an equivalence.

For general $\underline{A}$, the problem seems very difficult. It would, of course, be possible to continue forming categories $\underline{A}^{\mathbb{T}}$, $(\underline{A}^{\mathbb{T}})_{\mathbb{G}}$, ... Myles Tierney has given an example to show that this process needn't ever terminate. Presumably it is also possible that it terminate at any finite step. Thus we have, perhaps, a concept of a dimension of a category.

Here we give a complete answer when $\underline{A}$ is the category of sets (denoted by $\underline{S}$), pointed sets (denoted by $(1,\underline{S})$), or vector spaces over the field K (denoted by $\underline{V}_K$). Ignoring those $\mathbb{T}$ for which the functor T is constant, we show that $\dim\underline{S} = 1$ while $\dim(1,\underline{S}) = \dim\underline{V}_K = 0$. Since the main difference between $\underline{S}$ and $(1,\underline{S})$ is that the former contains some monomorphisms which do not split (any $\phi \longrightarrow X$), this concept of dimension does seem to be some kind of homological measure of the category. For the meaning and statements of the various tripleableness theorems we refer to [Be] and [Li].

1. PRELIMINARIES

According to the dual of the tripleableness theorem we must consider the question of whether F reflects isomorphisms and whether it preserves equalizers of F-split pairs (since all of our categories are complete.) But U reflects isomorphisms and creates all limits and an F-split pair is also UF-split. Thus we need only consider these questions for T itself. First we consider the question of T reflecting isomorphisms. In any concrete category we may call a triple consistent if it has a model of cardinality at least 2. We henceforth assume that $\underline{A}$ is one of the three categories mentioned above. The following lemma is from [La].

Lemma 1.

Let $\mathbb{T} = (T, \eta, \mu)$ be a consistent triple in $\underline{A}$. Then η is 1-1.

Proof. Any object $A \in \underline{A}$ with cardinality ≥ 2 is a cogenerator. Then for all $A' \in \underline{A}$, $A' \subset A^X$ for some set X. If A is a $\mathbb{T}$-algebra, so is A^X, and the given embedding $A' \hookrightarrow A^X$ can be completed to

so that $\eta A'$ is 1-1.

From now on we assume $\mathbb{T}$ is consistent. It is easily seen that on $\underline{S}$ there are 2 inconsistent triples ($TX = 1$ for all X is one and $TX = 1$ for all $X \neq \phi$, $T\phi = \phi$ is the other), while on $(1,\underline{S})$ and $\underline{V}_K$ there is exactly one.

<u>Proposition 2.</u>

T is faithful.

<u>Proof</u>. If $f \neq g: X \longrightarrow Y$, then since η is 1-1, $Tf.\eta X = \eta Y.f \neq \eta Y.g = Tg.\eta X$ so $Tf \neq Tg$.

<u>Theorem 3.</u>

T (and hence F) reflects isomorphisms.

<u>Proof.</u> For T, being faithful, reflects both epimorphisms and monomorphisms, and $\underline{A}$ has the property that a map which is both is an isomorphism.

2. SPLIT EQUALIZERS

In order to apply the cotripleableness theorem, it is necessary to have a combinatorial description of split equalizers. It turns out to be the same (almost) in all three categories.

Theorem 4.

Two maps $X \underset{d^1}{\overset{d^\circ}{\rightrightarrows}} Y$ can be put into a split equalizer diagram

$$E \underset{t}{\overset{d}{\rightleftarrows}} X \underset{s}{\overset{}{\rightrightarrows}} Y$$

satisfying $td = E$, $sd^\circ = X$, $sd^1 = dt$, $d^\circ d = d^1 d$ if the following conditions are satisfied.

(i) Equalizer $(d^\circ, d^1) \neq \phi$ (or $X = Y = \phi$).

(ii) d° is 1-1.

(iii) If $d^\circ x = d^1 x'$, then $d^\circ x = d^1 x$.

Condition (iii) may be described by saying that there is a map $u\colon P \longrightarrow E$ making the following diagram commute

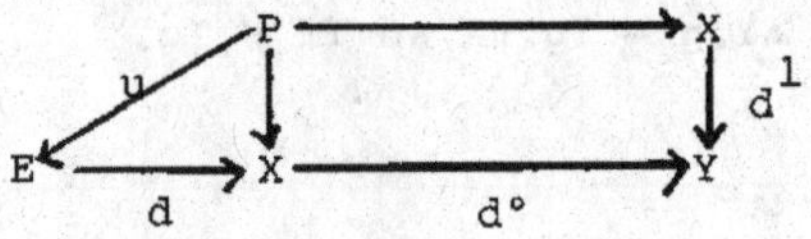

where P is the pullback (kernel pair) and E the equalizer.

Proof. The necessity of (i) and (ii) is clear (condition (i) refers only to S anyway.) As for (iii), if $d^\circ x = d^1 x'$, then $d^1 x = d^1 sd^\circ x = d^1 sd^1 x' = d^1 dtx' = d^\circ dtx' = d^\circ sd^1 x' = d^\circ sd^\circ x = d^\circ x$.

It is seen that these conditions will be necessary in any concrete category. Their sufficiency depends very heavily on the explicit categories at hand. We must consider cases. It is only necessary to find $s\colon Y \longrightarrow X$ with $sd^\circ = X$

and $d^\circ sd^1 = d^1sd^1$, for then $t: Y \longrightarrow E$ can be chosen so that $dt = sd^1$ by the nature of equalizers.

Case $\underline{S}$. Write $Y = Y_\circ + Y_1$ where $Y_\circ = \mathrm{Im}X$. Then $d^\circ: X \xrightarrow{\cong} Y_\circ$ (abusing notation). Choose $x_\circ \in X$ with $d^\circ x_\circ = d^1 x_\circ$. Then define $s: Y \longrightarrow X$ by $s \mid Y_\circ = (d^\circ)^{-1}$ and $s \mid Y_1$ is constant $x_\circ$. Clearly $sd^\circ = X$. If $d^1 x \notin \mathrm{Im}d^\circ$, then $sd^1x = x_\circ$ and $d^1sd^1x = d^\circ sd^1x$. If $d^1x = d^\circ x'$, then $d^\circ x' = d^1x'$, and then $d^1sd^1x = d^1sd^\circ x' = d^1x' = d^\circ x' = d^\circ sd^\circ x' = d^\circ sd^1x$.

Case $(1,\underline{S})$. This is exactly the same except that $x_\circ$ should be taken to be the basepoint.

Case $\underline{V}_K$. Let E be the equalizer of d° and d^1 and write $X = X_\circ + X_1$ where $d: E \xrightarrow{\cong} X_\circ$. We may assume $E = X_\circ$. Now since d° is 1-1, we can assume that $Y = X_\circ + X_1 + Y_2$ and d° is the inclusion. Moreover, $y \in \mathrm{Im}\, d^\circ \cap \mathrm{Im}\, d^1 \Longrightarrow y = d^\circ x = d^1x'$. Then $y = d^\circ x = d^1x$ and so $y \in \mathrm{Im}\, d^\circ d = X_\circ$. Hence we may choose Y_2 so that $\mathrm{Im}\, d^1 \subset X_\circ + Y_2$. In terms of this decomposition the maps d, d°, d^1 have matrix representations as follows.

$$d = \begin{pmatrix} X_\circ \\ \\ 0 \end{pmatrix}, \qquad d^\circ = \begin{pmatrix} X_\circ & 0 \\ 0 & X_1 \\ 0 & 0 \end{pmatrix}, \qquad d^1 = \begin{pmatrix} X_\circ & \alpha \\ 0 & 0 \\ 0 & \beta \end{pmatrix}$$

where $\alpha: X_1 \longrightarrow X_\circ$ and $\beta: X_1 \longrightarrow Y_2$ are arbitrary.

Let

$$s = \begin{pmatrix} X_\circ & 0 & 0 \\ 0 & X_1 & 0 \end{pmatrix}, \quad t = \begin{pmatrix} X_\circ, \alpha \end{pmatrix}$$

and then the required equations are clear.

3. PRESERVATION OF T-SPLIT EQUALIZERS

In this we show that T-split equalizers are preserved in cases $(1,\underline{S})$ and $\underline{V}_K$ and examine the sole failure of this in $\underline{S}$. The methods are really quite vulgar, since they prove that T (and hence F) is co-VTT.

Proposition 5

T (and hence F) reflects conditions (ii) and (iii) of theorem 4.

Proof. If $d^\circ : X \longrightarrow Y$ and Td° is 1-1, then $\mu Y.d^\circ = Td^\circ.\mu X$ and μX is 1-1, so $\mu Y.d^\circ$ and hence d° is 1-1. For the other part we use the diagram.

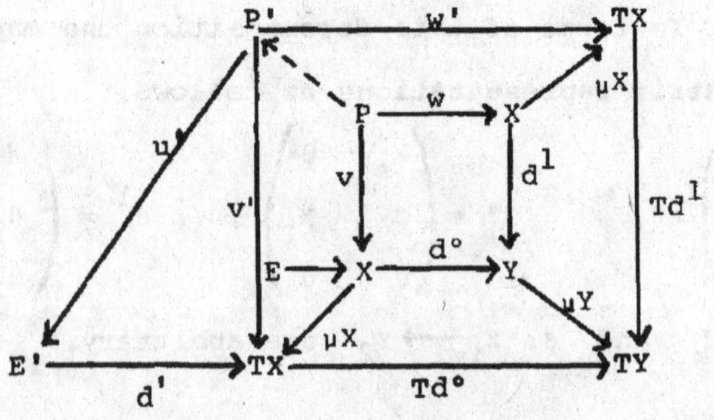

It is only necessary to show that $d^\circ v = d^1 v$. The universal property of P' allows a map $\pi: P \longrightarrow P'$ as indicated. Now $\mu Y.d^\circ.v = Td^\circ.\mu X.v = Td^\circ.v'.\pi = Td^\circ.d'.u'.\pi = Td^1.d'.u'.\pi = Td^1.v'.\pi = Td^1.\mu X.v = \mu Y.d^1.v$ and by lemma 1, μY is 1-1, so $d^\circ.v = d^1.v$.

<u>Theorem 6.</u>

In cases $(1,\underline{S})$ and $\underline{V}_K$, F is co-VTT.

<u>Proof.</u> There is nothing left to prove.

4. $\underline{S}$

There are easy examples which will become clear later which show that T does not necessarily reflect condition (i) of theorem 4.

<u>Theorem 7.</u>

If $T\phi \neq \phi$, then there is a factorization $(1,U): \underline{A}^{\mathbb{T}} \longrightarrow (1,\underline{S})$ such that

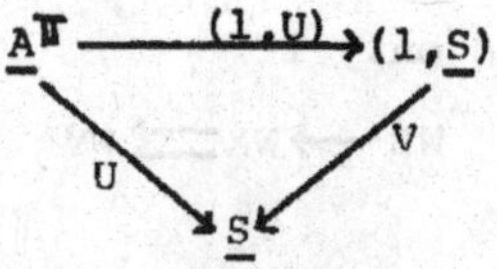

commutes, where V is the usual underlying functor. Moreover, there is a left adjoint $(1,F) \vdash (1,U)$ and $(1,F)$ is co-VTT.

Proof. Pick a point $a: 1 \longrightarrow T\phi$. Define $(1,U)A$ to be $1 \xrightarrow{a} T\phi = UF\phi \xrightarrow{U*} UA$ where $*: F\phi \longrightarrow A$ is the unique map in $(F\phi,A) = (\phi,UA)$. Clearly $V.(1,U) = U$. Define $(1,F)$ on objects by letting $(1,F)(1 \longrightarrow X) = F(X-\{1\})$. Then $(F(X-\{1\}),A) = (X-\{1\},UA) \approx (1,\underline{S})(1 \longrightarrow X, (1,U)A)$, the last isomorphism being obvious. Then $(1,F)$ has a unique extension to a functor left adjoint to $(1,U)$ and hence, by theorem 6, is co-VTT. Note that $1 \longrightarrow X \rightsquigarrow X-\{1\}$ cannot be extended to a functor, which is the reason for the indirect definition of $(1,F)$ first as an object function.

Theorem 8.

If $T\phi \neq \phi$, then F is cotripleable.

Proof. If $W: \underline{S} \longrightarrow (1,\underline{S})$ is defined by $WX = 1 \longrightarrow 1 + X$, then $W \vdash V$. But then $F = (1,F).W$, and since $(1,F)$ is co-VTT, it is only necessary to show that W is tripleable. It is, in fact, co-CTT, as may readily be checked. That is,

$$E \longrightarrow X \rightrightarrows Y$$

is an equalizer if and only if

$$WE \longrightarrow WX \rightrightarrows WY \qquad \text{is.}$$

Theorem 9.

Suppose e°, e^1 are the two distinct maps of $1 \longrightarrow 2$ (in $\underline{S}$). If $\phi \longrightarrow T1 \rightrightarrows T2$ is an equalizer, then F is co-tripleable.

Proof. It is sufficient, evidently, to show that if

$E' \xrightarrow{d'} TX \underset{Td^1}{\overset{Td^\circ}{\rightrightarrows}} TY$ is a split equalizer, then the equalizer

$$E \xrightarrow{d} X \underset{d^1}{\overset{d^\circ}{\rightrightarrows}} Y \qquad \text{is non-empty.}$$

At any rate condition (iii) is reflected, which means there will be a map $P \longrightarrow E$ so that if $E = \phi$, so is P. But if P is empty, this means we can find a map $2 \longrightarrow Y$ so that

$$\begin{array}{ccc} X & \underset{d^1}{\overset{d^\circ}{\rightrightarrows}} & Y \\ \downarrow & & \downarrow \\ 1 & \underset{e^1}{\overset{e^\circ}{\rightrightarrows}} & 2 \end{array}$$

commutes. This means each of the diagrams

$$\begin{array}{ccc} X & \xrightarrow{d^\circ} & Y \\ \downarrow & & \downarrow \\ 1 & \xrightarrow[e^\circ]{} & 2 \end{array} \qquad \begin{array}{ccc} X & \xrightarrow{d^1} & Y \\ \downarrow & & \downarrow \\ 1 & \xrightarrow[e^1]{} & 2 \end{array}$$

commutes. This provides us with a commutative diagram

$$\begin{array}{ccccc} E' & \longrightarrow & TX & \rightrightarrows & TY \\ \downarrow & & \downarrow & & \downarrow \\ \phi & \longrightarrow & T1 & \rightrightarrows & T2 \end{array}$$

so that $E' = \phi$. This completes the proof.

Now we suppose that $T\phi = \phi$, but that

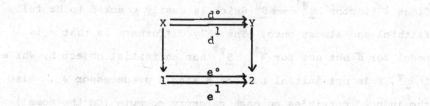

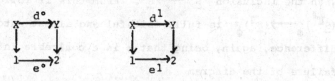

$\phi \neq E \longrightarrow T1 \underset{Te^1}{\overset{Te^\circ}{\rightrightarrows}} T2$ is an equalizer. Since the underlying functor creates equalizers, it is clear that E has the structure of a $\mathbb{T}$ algebra, so that there is a map $TE \xrightarrow{\mu^{\#}\phi} E$ defining this structure. Also we let $\eta^{\#}\phi\colon \phi \longrightarrow E$ denote the unique map. Let $T^{\#}$ be defined by $T^{\#}X = \begin{cases} TX & \text{if} \quad X \neq \phi \\ E & \text{if} \quad X = \phi \end{cases}$. It is trivial to check that with $\eta^{\#}X = \eta X$ and $\mu^{\#}X = \mu X$ for $X \neq \phi$, $\mathbb{T}^{\#} = (T^{\#}, \eta^{\#}, \mu^{\#})$ becomes a triple on $\underline{S}$. Moreover, $T^{\#}\phi \neq \phi$. Clearly every $\mathbb{T}^{\#}$-algebra is also a $\mathbb{T}$-algebra, since E is the only new free algebra and it was already an algebra. This defines a functor $\underline{S}^{\mathbb{T}^{\#}} \longrightarrow \underline{S}^{\mathbb{T}}$ which is easily checked to be full, faithful and almost onto. The only difference is that ϕ is a model for $\mathbb{T}$ but not for $\mathbb{T}^{\#}$. $\underline{S}^{\mathbb{T}^{\#}}$ has an initial object E, while in $\underline{S}^{\mathbb{T}}$, E is not initial but has a single predecessor ϕ. Also the induced cotriples on each category commute (on the nose!) with the inclusion $\underline{S}^{\mathbb{T}^{\#}} \longrightarrow \underline{S}^{\mathbb{T}}$. From this it follows that $(\underline{S}^{\mathbb{T}^{\#}})_{\mathbb{G}} \longrightarrow (\underline{S}^{\mathbb{T}})_{\mathbb{G}}$ is full, faithful and almost onto, the only difference, again, being that ϕ is a coalgebra in $(\underline{S}^{\mathbb{T}})_{\mathbb{G}}$. The failure of the diagram

$$\begin{array}{ccc} \underline{S}^{\mathbb{T}^{\#}} & \longrightarrow & \underline{S}^{\mathbb{T}} \\ & {\scriptstyle F^{\#}} \nwarrow \quad \nearrow {\scriptstyle F} & \\ & \underline{S} & \end{array}$$

to commute exactly — $F^{\#}\phi = E$, $F\phi = \phi$ — also makes

$$\begin{array}{ccc} (\underline{S}^{\mathbb{T}^{\#}})_{\mathbb{G}} & \longrightarrow & (\underline{S}^{\mathbb{T}})_{\mathbb{G}} \\ & {\scriptstyle \approx} \nwarrow \quad \nearrow {\scriptstyle \Psi} & \\ & \underline{S} & \end{array}$$

not quite commute. In fact the comparison Ψ is such that $\Psi\phi = \phi$. Thus $(\underline{S}^{\mathbb{T}})_{\mathbb{G}}$ is as in the following.

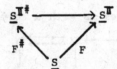

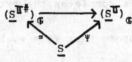

Theorem 10.

Suppose $T\phi = \phi$, but the equalizer $E \longrightarrow T1 \rightrightarrows T2$ is non-empty. Then $(\underline{S}^{\mathbb{T}})_{\mathbb{G}}$ may be described as $\underline{S} \cup \{\gamma\}$ where $(\phi,\gamma) = (\gamma,\gamma) = 1$ and $(X,\gamma) = \phi$ for any other $X \in \underline{S}$. The comparison $\underline{S} \xrightarrow{\Psi} (\underline{S}^{\mathbb{T}})_{\mathbb{G}}$ is just the inclusion functor. Moreover, this induced triple on $\underline{S} \cup \{\gamma\}$ has $\underline{S}^{\mathbb{T}}$ as its algebras.

Note that if $\mathbb{T}^{\#}$ is a triple with $T^{\#}\phi \neq \phi$, we may define $\mathbb{T}$ by $T\phi = \phi$ and get a 1-1 correspondence between triples for which $T\phi = \phi$, $E \neq \phi$ and these for which $T\phi \neq \phi$ (in which case it is easily seen-- owing to the lack of empty models-- that $T\phi = E$). On the other hand, if $E = \phi$ (as when $\mathbb{T} = \mathbb{P}$), this can't happen. E is called the set of pseudo-constants. This is justified by the fact that it is the set of constant natural transformations of $U \longrightarrow U$ (whereas $T\phi$ is the set of natural transformations of $U^{\phi} \longrightarrow U$ -- the indeterminancy of 0° rears its ugly head again). Then we may restate our results.

Theorem 11.

If the set of pseudo-constants of $\mathbb{T}$ is equal to the set of constants, then $(\underline{S}^{\mathbb{T}})_{\mathbb{G}} = \underline{S}$. Otherwise it is $\underline{S} \cup \{\gamma\}$ as above.

REFERENCES

[Be] J. Beck, "The Tripleableness Theorems," (to appear).

[La] F. W. Lawvere, "Functorial Semantics of Algebraic Theories", thesis, Columbia University, (1963).

[Li] F. E. J. Linton, "Applied Functorial Semantics, II," (to appear in *Zurich Lecture Notes*).

LECTURES ON REGULAR LOCAL RINGS

by D.A. Buchsbaum

§1. Koszul complexes

The aim of these lectures is to show how homological methods may be applied to local ring theory. That homology is tied up naturally with algebra, can be seen easily by looking at the following example: Let R be a commutative ring (all rings are assumed to have an identity element), and let x be an element of R. Then we may consider the rather trivial complex

$$\cdots \longrightarrow 0 \longrightarrow 0 \longrightarrow 0 \longrightarrow R \xrightarrow{x} R \longrightarrow 0$$

where $R \xrightarrow{x} R$ means multiplication by the element x. If we call this complex $K(x)$, we see that $H_0(K(x)) = R/(x)$, and $H_1(K(x)) = 0:x = \{r \varepsilon R / rx = 0\}$. Thus in a sense H_0 measures how divisible R is by x, while H_1 tells us whether or not x is a zero divisor. For an arbitrary R-module M, we may also consider the complex $K(x) \otimes_R M$:

$$\cdots \longrightarrow 0 \longrightarrow 0 \longrightarrow M \xrightarrow{x} M \longrightarrow 0$$

and observe that $H_0(K(x) \otimes M) = M/xM$ and $H_1(K(x) \otimes M) = 0:x = \{m \varepsilon M / xm = 0\}$. Again we may use these homology groups to tell us how divisible the module M is by the element x, and whether or not x is a zero divisor for M.

Now let us suppose that we have two elements x, y in R. Then we have the map $K(x) \xrightarrow{y} K(x)$ induced by multiplication by y:

$$\begin{array}{ccccccccccc} \cdots & \longrightarrow & 0 & \longrightarrow & 0 & \longrightarrow & R & \xrightarrow{x} & R & \longrightarrow & 0 \\ & & & & & & \downarrow y & & \downarrow y & & \\ \cdots & \longrightarrow & 0 & \longrightarrow & 0 & \longrightarrow & R & \xrightarrow{x} & R & \longrightarrow & 0 \end{array}$$

and it is the most natural thing in the world for homologists to consider the mapping cylinder of this map of complexes:

$$\cdots \longrightarrow 0 \longrightarrow 0 \longrightarrow R \xrightarrow{f} R \oplus R \xrightarrow{g} R \longrightarrow 0$$

where $g(r_1, r_2) = r_1 x + r_2 y$ and $f(r) = (-ry, rx)$. Calling this complex $K(x,y)$, we have $H_0(K(x,y)) = R/(x,y)$; $H_2(K(x,y)) = 0:(x,y) = \{r \varepsilon R / rx = 0 = ry\}$. $H_1(K(x,y)) = \{(r_1, r_2) / r_1 x + r_2 y = 0\} / \{(-ry, rx)\}$.

If we suppose that R is an integral domain, and that neither x nor y is zero, we see that $H_2(K(x,y)) = 0$ and that $\{(r_1,r_2)/r_1x+r_2y = 0\}$ is mapped isomorphically onto $x:y = \{r\varepsilon R/ry\varepsilon(x)\}$ by the map sending $(r_1,r_2) \longrightarrow r_2$. Under this map, the submodule $\{(-ry,rx)\}$ is sent onto the ideal (x) and thus $H_1(K(x,y)) = (x:y)/(x)$. If R were a unique factorization domain, then $H_1(K(x,y))$ would be zero if and only if x and y were relatively prime. Therefore we again see that a certain amount of homological information is intimately connected with arithmetical questions.

We could proceed step by step in this way, building up complexes $K(x,y,z,\ldots)$, but it is more efficient to consider the following general construction. Let $f:A \longrightarrow R$ be a morphism from the R-module A to R. Then we may define a complex $K(f)$ by setting $(K(f))_p = \overset{p}{\Lambda}A$ and defining $df:\overset{p}{\Lambda}A \longrightarrow \overset{p-1}{\Lambda}A$ by $df(a_1\wedge\ldots\wedge a_p) = \Sigma(-1)^{i+1}f(a_i)a_1\wedge\ldots\wedge\hat{a}_i\wedge\ldots\wedge a_p$ where $\hat{a}_i$ means we omit a_i. It is well-known, and easy to check, that $df^2 = 0$. Also, if $\alpha\varepsilon\overset{p}{\Lambda}A$ and $\beta\varepsilon\overset{q}{\Lambda}A$, then $df(\alpha\wedge\beta) = df(\alpha)\wedge\beta+(-1)^p\alpha\wedge df(\beta)$. Before indicating some general properties of this complex, we should see how it subsumes the sort of thing we have been discussing. Namely, if $x_1,\ldots,x_s$ is a sequence of elements of our ring R, then we can define a morphism $f:R^s \longrightarrow R$ where R^s denotes the direct sum of R with itself s times, and f is defined by $f(1,0,\ldots,0) = x_1,\ldots,f(0,0,\ldots,1) = x_s$. In this case, we denote the complex $K(f)$ by $K(x_1,\ldots,x_s)$.

Returning to the general case of a morphism $f:A \longrightarrow R$, we state two fundamental properties of the complex $K(f)$:

<u>Proposition 1.1</u>. If a_0 is an element of A, then $f(a_0)H_p(K(f)) = 0$ for all $p \geq 0$.

<u>Proposition 1.2</u>. If x is any element of R, and $f:A \longrightarrow R$ a morphism, we have the map of complexes $x:K(f) \longrightarrow K(f)$:

$$\begin{array}{ccccccccc}
\ldots & \longrightarrow & \overset{3}{\Lambda}A & \longrightarrow & \overset{2}{\Lambda}A & \longrightarrow & A & \xrightarrow{f} & R \\
 & & \downarrow x & & \downarrow x & & \downarrow x & & \downarrow x \\
\ldots & \longrightarrow & \overset{3}{\Lambda}A & \longrightarrow & \overset{2}{\Lambda}A & \longrightarrow & A & \longrightarrow & R
\end{array}$$

The mapping cylinder of this map of complexes is $K(\overline{f})$ where $\overline{f}:A\oplus R \longrightarrow R$ is defined by $\overline{f}((a,r)) = f(a)+rx$.

The proof of Proposition 1 is obtained by writing a homotopy $s:\overset{p}{\Lambda}A \longrightarrow \overset{p+1}{\Lambda}A$ defined by $s(a_1\wedge\ldots\wedge a_p) = a_0\wedge a_1\wedge\ldots\wedge a_p$, and showing that multiplication by $f(a_0)$ is chain homotopic to the zero morphism. Proposition 2 is proved by observing that for any module A, $\overset{p}{\Lambda}(A\oplus R) \approx \overset{p}{\Lambda}A \oplus \overset{p-1}{\Lambda} A$.

From Proposition 2, we obtain

Corollary 1.3. If $f:A \longrightarrow R$ is a morphism, x an element of R, and $\bar{f}:A\oplus R \longrightarrow R$ defined as in Proposition 2, we have the exact sequence:

$$\cdots \longrightarrow H_p(K(f)) \longrightarrow H_p(K(\bar{f})) \longrightarrow H_{p-1}(K(f)) \xrightarrow{\pm x} H_{p-1}(K(f)) \longrightarrow \cdots \longrightarrow H_0(K(f)) \longrightarrow H_0(K(\bar{f})) \longrightarrow 0 .$$

In particular, if $x_1,\ldots,x_{s+1}$ is a sequence of elements of R, we have the exact sequence:

$$(E): \cdots \to H_p(K(x_1,\ldots,x_s)) \to H_p(K(x_1,\ldots,x_{s+1})) \to H_{p-1}(K(x_1,\ldots,x_s)) \xrightarrow{\pm x_{s+1}} H_{p-1}(K(x_1,\ldots,x_s)) \to \cdots$$

Using the fact that $K(x_1,\ldots,x_s)$ and $K(x_1,\ldots,x_{s+1})$ are free chain complexes, we also get, for an arbitrary R-module M, the exact sequence:

$$(E(M)): \cdots \to H_p(M(x_1,\ldots,x_s)) \to H_p(M(x_1,\ldots,x_{s+1})) \to H_{p-1}(M(x_1,\ldots,x_s) \xrightarrow{\pm x_{s+1}} H_{p-1}(M(x_1,\ldots,x_s) \to \cdots$$

where $M(x_1,\ldots,x_s)$ is the complex $K(x_1,\ldots,x_s)\otimes_R M$.

From the remarks made about the meaning of $H_1(K(x,y))$, we are led to make the following definition.

Definition: Let M be an R-module, and $x_1,\ldots,x_s$ a sequence of elements of R. Then $x_1,\ldots,x_s$ is said to be an M-sequence if a) $M/(x_1,\ldots,x_s)M \neq 0$ and b) for each i, $1 \leq i \leq s$, the element x_i is not a zero divisor for $M/(x_1,\ldots,x_{i-1})M$. For $i=1$, this means that x_1 is not a zero divisor for M.

Proposition 1.4. If $x_1,\ldots,x_s$ is an M-sequence, then $H_p(M(x_1,\ldots,x_s)) = 0$ for all $p > 0$.

The proof proceeds by induction on s, the case $s=1$ being self-evident. To go from s to $s+1$, one merely uses the exact sequence $E(M)$ to show that $H_p(M(x_1,\ldots,x_s)) = 0$ for $p > 1$, and the fact that $H_0(M(x_1,\ldots,x_{s-1})) \xrightarrow{x_s} H_0(M(x_1,\ldots,x_{s-1}))$ is a monomorphism (since $H_0(M(x_1,\ldots,x_{s-1})) = M/(x_1,\ldots,x_{s-1})M)$ to show that $H_1(M(x_1,\ldots,x_s)) = 0$.

Remark: It is not true in general that $H_p(M(x_1,\ldots,x_s)) = 0$ for all $p > 0$ implies that $x_1,\ldots,x_s$

is a M-sequence. For example, let $R = k[x,y,x]/y(x-1,z)$, with k a field. If we write $\bar{x}$ and $\bar{z}$ for the residue classes of x and z in R, we see easily that $\bar{x},\bar{z}$ is an R-sequence, so that $H_p(K(\bar{x},\bar{z})) = 0$ for $p > 0$. Therefore, we obviously have that $H_p(K(\bar{z},\bar{x})) = 0$ (since for any ring R, and any sequence of elements $x_1,\ldots,x_s$ in R, and any R-module M, we have $M(x_1,\ldots,x_s) \approx M(x_{\pi(1)},\ldots,x_{\pi(s)})$ where π is a permutation of the set $\{1,\ldots,s\}$). However, it is clear that $\bar{z}$ is a zero divisor in R, so that $\bar{z},\bar{x}$ fails to be an R-sequence.

As a result of the above remark, it is natural to ask when we can prove the converse of Proposition 4. To this end, let us suppose that R is a noetherian ring, M a finitely generated R-module, and $x_1,\ldots,x_s$ elements in the radical of R. Recall that the radical of R is the intersection of all the maximal ideals of R. We now have Nakayama's lemma at our disposal, which says:

Lemma 1.5. Let R be a ring, M a finitely generated R-module, and $\mathfrak{a}$ an ideal in the radical of R such that $\mathfrak{a}M = M$. Then $M = 0$.

If we now use our assumptions that R is noetherian and M finitely generated, we conclude that $H_p(M(x_1,\ldots,x_s)$ is a finitely generated R-module for any sequence of elements $x_1,\ldots,x_s$ in R. Moreover, if x_i is in the radical of R, we can then conclude that the map $H_p(M(x_1,\ldots,\hat{x}_i,\ldots,x_s)) \xrightarrow{x_i} H_p(M(x_1,\ldots,\hat{x}_i,\ldots,x_s))$ is an epimorphism if and only if $H_p(M(x_1,\ldots,\hat{x}_i,\ldots,x_s)) = 0$. Using these facts, we can prove

Proposition 1.6. Let R be a noetherian ring, M a finitely generated R-module, and $x_1,\ldots,x_s$ a sequence of elements in the radical of R. Then

i) if $H_p(M(x_1,\ldots,x_s)) = 0$ for some p, we have $H_{p+k}(M(x_1,\ldots,x_t)) = 0$ for all $k \geq 0$ and all t, $1 \leq t \leq s$;

ii) if $H_1(M(x_1,\ldots,x_s)) = 0$, then $x_1,\ldots,x_s$ is an M-sequence. Thus, under these hypotheses, we have $x_1,\ldots,x_s$ is an M-sequence if and only if $H_p(M(x_1,\ldots,x_s)) = 0$ for all $p > 0$ (or for $p = 1$).

The proofs of these statements are obtained by repeated use of the exact sequence E(M), together with the observations immediately preceding the Proposition.

Corollary 1.7. The hypotheses being as in Proposition 6, we have $x_1,\ldots,x_s$ is an M-sequence if and only if $x_{\pi(1)},\ldots,x_{\pi(s)}$ is an M-sequence for every permutation π of $\{1,\ldots,s\}$.

Corollary 1.8. If R is a noetherian ring, M a finitely generated R-module, $\mathfrak{p}$ a prime ideal of R

such that $M_{\mathfrak{p}} = M \otimes_R R_{\mathfrak{p}} \neq 0$ (i.e. $\mathfrak{p}$ is in Supp (M)), and $x_1, \ldots, x_s$ a sequence of elements in $\mathfrak{p}$ which is an M-sequence, then $x_1, \ldots, x_s$, considered as elements in $R_{\mathfrak{p}}$, is an M -sequence.

Remarks: In general, if R is any commutative ring, and S a multiplicatively stable subset of R, we denote by R_S the ring of quotients of R with respect to S. However, if $\mathfrak{p}$ is a prime ideal of R, and $S = R - \mathfrak{p}$, we denote the ring of quotients R_S by $R_{\mathfrak{p}}$. If R is noetherian, then R_S is noetherian. If R is noetherian and $\mathfrak{p}$ a prime ideal, then $R_{\mathfrak{p}}$ is a local ring i.e. a noetherian ring with one maximal ideal. In the case of $R_{\mathfrak{p}}$, the unique maximal ideal is $\mathfrak{p}R_{\mathfrak{p}}$. Finally, R_S is flat over R for any multiplicatively stable subset S of R.

§2. Local rings

We shall now assume that R is a local ring, with maximal ideal $\mathfrak{m}$. The residue field $R/\mathfrak{m}$ will be denoted by k; all R-modules will be assumed finitely generated. In this case, the radical of R is $\mathfrak{m}$, and all proper ideals of R are contained in $\mathfrak{m}$. If E is an R-module and $\mathfrak{a}$ an ideal of R, we have $E/\mathfrak{a}E = 0$ if and only if $E = 0$ because of Nakayama's Lemma.

Lemma 2.1. Let E be an R-module (always finitely generated), and $e_1, \ldots, e_t$ elements of E such that $\bar{e}_1, \ldots, \bar{e}_t$ generate $E/\mathfrak{m}E$ as a k-vector space. Then $e_1, \ldots, e_t$ generate E as an R-module.

Proof. Although the proof of this is well-known, we will reproduce it here since it serves as a prototype for so many others. We take a free module F of rank t and map it onto the submodule of E generated by $e_1, \ldots, e_t$. We then have the exact sequence $F \longrightarrow E \longrightarrow L \longrightarrow 0$ where L is the cokernel of $F \longrightarrow E$. Tensoring over R with k, we have the exact sequence $F/\mathfrak{m}F \longrightarrow E/\mathfrak{m}E \longrightarrow L/\mathfrak{m}L \longrightarrow 0$. However, $F/\mathfrak{m}F \longrightarrow E/\mathfrak{m}E$ is now an epimorphism, so $L/\mathfrak{m}L = 0$. This tells us that $L = 0$ so that $F \longrightarrow E \longrightarrow 0$ is exact and $e_1, \ldots, e_t$ generate E.

Corollary 2.2. Any generating set of an R-module E contains a minimal generating set of E; any two minimal generating sets of E have the same number of elements. This number is equal to $[E/\mathfrak{m}E:k]$ i.e. the dimension of the vector-space $E/\mathfrak{m}E$ over k. Moreover, any subset of E linearly independent modulo $\mathfrak{m}E$ may be extended to a minimal generating set of E.

Lemma 2.3. If E is an R-module, there exists a free module F and an epimorphism $g: F \longrightarrow E$ such that Ker g is contained in $\mathfrak{m}F$.

Lemma 2.4. If E is an R-module such that $\mathrm{Tor}_1^R(k,E) = 0$, then E is free.

Lemma 2.5. For every R-module E, $\mathrm{hd}_R E \leq \mathrm{hd}_R k$.

Proposition 2.6. If R is a local ring, $\mathrm{gl.dim}\, R = \mathrm{hd}_R k$.

The proofs of all the above statements are completely straightforward. For instance, 2.3 depends on nothing deeper than the fact that an epimorphism of a vector space onto another of the same dimension is an isomorphism. One then chooses F to be free on the number of generators in a minimal generating set of E.

The main result we are heading for now is that $[\mathfrak{m}/\mathfrak{m}^2 : k] \leq \mathrm{gl.dim}\, R$, i.e. that the global dimension of R is never less than the minimal number of generators of the maximal ideal of R. The idea of the proof is the following

We take a minimal generating set $x_1, \ldots, x_n$ of the maximal ideal $\mathfrak{m}$ and form the complex $K(x_1, \ldots, x_n)$. We then show that we can find a free resolution of k:

$$\cdots \longrightarrow X_3 \xrightarrow{d_3} X_2 \xrightarrow{d_2} X_1 \xrightarrow{d_1} R \longrightarrow k \longrightarrow 0$$

such that $d_i(X_i) \subset \mathfrak{m}X_{i-1}$ and such that $K(x_1, \ldots, x_n)$ is a subcomplex of this resolution. Tensoring this resolution with k makes the boundary maps zero, so that $\mathrm{Tor}_p^R(k,k) = X_p/\mathfrak{m}X_p$. Since $K(x_1, \ldots, x_n)$ is a subcomplex of the resolution, and $K(x_1, \ldots, x_n)$ is non-zero in dimension n, we see that $\mathrm{Tor}_n^R(k,k) \neq 0$. Thus $\mathrm{hd}_R k \geq n$ and we have the desired result using the facts that $\mathrm{gl.dim}\, R = \mathrm{hd}_R k$ and $n = [\mathfrak{m}/\mathfrak{m}^2 : k]$.

The complex $K(x_1, \ldots, x_n)$ is of the form:

$$0 \longrightarrow \overset{n}{\Lambda} R^n \xrightarrow{\delta_n} \cdots \longrightarrow \overset{2}{\Lambda} R^n \xrightarrow{\delta_2} R^n \xrightarrow{f} R .$$

We may therefore choose X_1 of our resolution to be R^n with $d_1 : X_1 \longrightarrow R$ defined to be the morphism f. Assume now that $X_1, \ldots, X_p$ have been found such that i) $X_k = X_k' \oplus \overset{k}{\Lambda} R^n$: ii) the morphism $d_k : X_k \longrightarrow X_{k-1}$ restricted to $\overset{k}{\Lambda} R^n$ is δ_k; iii) $\mathrm{Ker}\, d_k \subset \mathfrak{m}X_k$ and iv)

iv) $X_p \longrightarrow \cdots \longrightarrow X_1 \longrightarrow R \longrightarrow k \longrightarrow 0$ is exact. If $p < n$, we will show that we can find a free module X_{p+1} such that i) - iv) are true with k replaced by $p + 1$. Our conditions i) and ii) tell us that the composition $\Lambda^{p+1} R^n \xrightarrow{\delta_{p+1}} \Lambda^{p} R^n \longrightarrow X_p \xrightarrow{d_p} X_{p-1}$ is zero. Thus $\delta_{p+1}(\Lambda^{p+1} R^n)$ is contained in $Z_p = \text{Ker}\, d_p$. We will show that if $\{\varepsilon_{i_1} \wedge \dots \wedge \varepsilon_{i_{p+1}}\}$ are a basis of $\Lambda^{p+1} R^n$, then $\{\delta_{p+1}(\varepsilon_{i_1} \wedge \dots \wedge \varepsilon_{i_{p+1}})\}$ are linearly independent modulo $\mathfrak{m}Z_p$. That is, we want to show that if $\Sigma\, \gamma_{i_1 \dots i_{p+1}}\, \delta_{p+1}(\varepsilon_{i_1} \wedge \dots \wedge \varepsilon_{i_{p+1}}) \in \mathfrak{m}Z_p$, then each $\gamma_{i_1 \dots i_{p+1}}$ is in $\mathfrak{m}$. But this is the same as showing that if $\;\; = \Sigma\, \gamma_{i_1 \dots i_{p+1}}\, \varepsilon_{i_1} \wedge \dots \wedge \varepsilon_{i_{p+1}}$ in $\Lambda^{p+1} R^n$ is such that $\delta_{p+1}(\;\;)$ is in $\mathfrak{m}Z_p$, then is in $\mathfrak{m}\Lambda^{p+1} R^n$. Suppose, then, that $\delta_{p+1}(\;\;)$ is in $\mathfrak{m}Z_p$. Since $Z_p \subset \mathfrak{m}X_p$, we have $\mathfrak{m}Z_p \subset \mathfrak{m}^2 X_p = \mathfrak{m}^2 \Lambda^p R^n \oplus \mathfrak{m}^2 X'_p$. Since $\delta_{p+1}(\;\;)$ is in $\Lambda^p R^n$, we must have $\delta_{p+1}(\;\;) \in \mathfrak{m}^2 \Lambda^p R^n$.

Now $\delta_{p+1}: \Lambda^{p+1} R^n \longrightarrow \Lambda^p R^n$ induces a map $\bar{\delta}_{p+1}: \Lambda^{p+1} R^n / \mathfrak{m} \Lambda^{p+1} R^n \longrightarrow \mathfrak{m}\Lambda^p R^n / \mathfrak{m}^2 \Lambda^p R^n$ since $\delta_{p+1}(\Lambda^{p+1} R^n) \subset \mathfrak{m}\Lambda^p R^n$. If we can show that $\bar{\delta}_{p+1}$ is a monomorphism, we are done. For to say that $\delta_{p+1}(\gamma)$ is in $\mathfrak{m}^2 \Lambda^p R^n$ is to say that $\bar{\delta}_{p+1}(\bar{\gamma}) = 0$ and thus that $\bar{\gamma} = 0$ or $\gamma \in \mathfrak{m}\Lambda^{p+1} R^n$. But $\Lambda^{p+1} R^n / \mathfrak{m} \Lambda^{p+1} R^n = R/\mathfrak{m} \otimes \Lambda^{p+1} R^n$, and $\mathfrak{m}\Lambda^p R^n / \mathfrak{m}^2 \Lambda^p R^n = \mathfrak{m}/\mathfrak{m}^2 \otimes \Lambda^p R^n$ and the map δ_{p+1} sends the basis element $1 \otimes \varepsilon_{i_1} \wedge \dots \wedge \varepsilon_{i_{p+1}}$ to $\Sigma(-1)^{j+1} \bar{x}_{i_j} \otimes \varepsilon_{i_1} \wedge \dots \wedge \hat{\varepsilon}_{i_j} \wedge \dots \wedge \varepsilon_{i_{p+1}}$. Clearly this map is a monomorphism, and we are done.

Since we have shown that $\{\delta_{p+1}(\varepsilon_{i_1} \wedge \dots \wedge \varepsilon_{i_{p+1}})\}$ are linearly independent modulo $\mathfrak{m}Z_p$, we know that these elements may be chosen as part of a minimal generating set for Z_p, i.e. we have a minimal generating set $\{\delta_{p+1}(\varepsilon_{i_1} \wedge \dots \wedge \varepsilon_{i_{p+1}})\} \cup \{z_1, \dots, z_q\}$. Letting X'_{p+1} be the free module on q generators, we have $\Lambda^{p+1} R^n \oplus X'_{p+1} = X_{p+1}$ mapping onto Z_p. The kernel of the map $X_{p+1} \longrightarrow Z_p \longrightarrow X_p$ is clearly in $\mathfrak{m}X_{p+1}$ and $X_{p+1} \longrightarrow X_p \longrightarrow \dots \longrightarrow X_1 \longrightarrow R \longrightarrow k \longrightarrow 0$ is exact. Thus we have completed the inductive step, and we have proved

Theorem 2.6. If R is a local ring, then $[\mathfrak{m}/\mathfrak{m}^2 : k] \leq \text{gl.dim}\, R$.

Our next objective is to define the dimension (or Krull dimension) of a local ring R, and show that $\dim R \leq [\mathfrak{m}/\mathfrak{m}^2 : k]$. To do this, we shall use the method of Hilbert-Samuel polynomials which we introduce in a slightly more general setting now.

§3. Hilbert-Samuel Polynomials.

We let Z^+ denote the set of positive integers, and let G be an abelian group. Generally, G will be the group of integers or the Grothendieck group of some category of modules.

Definitions. Let $f:Z^+ \longrightarrow G$ be a function. f is called a polynomial function if there are elements $a_0,\ldots,a_d$ in G such that for all $n \in Z^+$, n sufficiently large (i.e. $n \gg 0$), we have $f(n) = \sum_{i=0}^{d} \binom{n}{i} a_i$ where $\binom{n}{i}$ denotes the binomial coefficient $\frac{n!}{i!(n-i)!}$. If $f:Z^+ \longrightarrow G$ is any function, we define $\Delta f:Z^+ \longrightarrow G$ by $\Delta f(n) = f(n+1)-f(n)$. Δf is called the first difference (function) of f. For any integer $s > 1$, we define $\Delta^s f = \Delta(\Delta^{s-1} f)$.

Lemma 3.1. A function $f:Z^+ \longrightarrow G$ is a polynomial function if and only if Δf is a polynomial function.

Proof. If f is a polynomial function, we have $f(n) = \sum_{i=0}^{d} \binom{n}{i} a_i$ for some $a_0,\ldots,a_d$ in G, and all $n \gg 0$. Consequently, for all $n \gg 0$ we have $\Delta f(n) = f(n+1)-f(n) = \Sigma\binom{n+1}{i}a_i - \Sigma\binom{n}{i}a_i =$ $\Sigma[\binom{n+1}{i}-\binom{n}{i}]a_i = \sum_{i=1}^{d} \binom{n}{i-1} a_i$. Thus Δf is a polynomial function.

Conversely, if $\Delta f(n) = \sum_{i=0}^{d} \binom{n}{i} b_i$ for $n \gg 0$, define $g(n) = f(n) - \sum_{i=0}^{d} \binom{n}{i+1} b_i$. Then for $n \gg 0$, $\Delta g(n) = 0$ so that $g(n)$ is a constant a_0. Thus, letting $a_i = b_{i-1}$ for $i > 0$, we have $f(n) = \sum_{i=0}^{d+1} \binom{n}{i} a_i$ for $n \gg 0$.

We have proved Lemma 3.1 in detail, since the definition of polynomial function is slightly more general than the usual one. However, with 3.1 established, we shall only state the facts we need, and omit the proofs.

Lemma 3.2. If $f(n) = \sum_{i=0}^{d} \binom{n}{i} a_i = \sum_{i=0}^{d'} \binom{n}{i} a'_i$ is a polynomial function, then $d = d'$ and $a_i = a'_i$. Thus, the degree, d, of f is a well defined integer, and the coefficients a_i of f are uniquely determined.

Now consider a commutative noetherian ring R, and let $\mathcal{A}$ be a full abelian subcategory of the category of R-modules. For each integer $s = 0,1,2,\ldots$, let $\mathcal{A}_s$ be the category of finitely generated graded $R[X_1,\ldots,X_s]$-modules $E = \sum_{\nu\geq 0} E_\nu$ such that

i) if $s = 0$, E_ν is in $\mathcal{A}$ for all ν;

ii) if $s > 0$, E_ν is in $\mathcal{A}$ for all ν and the graded $R[X_1,\ldots,X_{s-1}]$ modules $\mathrm{Ker}\,(\Sigma E_\nu \xrightarrow{X_s} \Sigma E_\nu)$ and $\mathrm{Coker}\,(\Sigma E_\nu \xrightarrow{X_s} \Sigma E_\nu)$ are in $\mathcal{A}_{s-1}$.

Finally, let f_0 be a function from the objects of $\mathcal{A}$ to an abelian group G which factors through the Grothendieck group, $K(\mathcal{A})$, of i.e. f_0 is additive with respect to exact sequences in $\mathcal{A}$. Then

<u>Theorem 3.3</u>. Let $E = \Sigma E_\nu$ be an object in $\mathcal{A}_s$ and define $f_E : Z^+ \longrightarrow G$ by $f_E(\nu) = f_0(E_\nu)$. Then f_E is a polynomial function of degree less than or equal to $s-1$.

An example of such a set-up occurs when R is a local ring, and $\mathcal{A}$ is the full subcategory of R-modules of finite length. In that case, an $R[X_1,\ldots,X_s]$-module $E = \Sigma_\nu$ is in $\mathcal{A}_s$ if it is finitely generated, and if each E_ν is an R-module of finite length. We may then choose G to be the group, Z, of integers and $f_0(E)$ = length of E. This is of course the most usual example. To see that it is not the only example, we may choose $\mathcal{A}$ to be the category of all finitely generated R-modules, and G to be $K(\mathcal{A})$ itself with f_0 the usual map.

<u>Corollary 3.4</u>. If R is a local ring, E a finitely generated R-module and $\mathfrak{q}$ an ideal of R containing some power $\mathfrak{m}^n$ of the maximal ideal $\mathfrak{m}$, then $E/\mathfrak{q}^\nu E$ is an R-module of finite length, and the function $\chi_{\mathfrak{q}}(E) : Z^+ \longrightarrow Z$ defined by $\chi_{\mathfrak{q}}(E;\nu)$ = length of $E/\mathfrak{q}^\nu E$ is a polynomial function whose degree is less than or equal to $[\mathfrak{q}/\mathfrak{m}\mathfrak{q}:k]$.

To see why this is a corollary of 3.3 we observe that we have exact sequences

$$0 \longrightarrow {}^\nu E/\mathfrak{q}^{\nu+1}E \longrightarrow E/\mathfrak{q}^{\nu+1}E \longrightarrow E/\mathfrak{q}^\nu E \longrightarrow 0$$

and therefore $\Delta\chi(E;\nu)$ = length of $\mathfrak{q}^\nu E/\mathfrak{q}^{\nu+1}E$. The graded module $\Sigma\mathfrak{q}^\nu E/\mathfrak{q}^{\nu+1}E$ is a module over $R[X_1,\ldots,X_s]$ where $s = [\mathfrak{q}/\mathfrak{m}:k]$, and each $\mathfrak{q}^\nu E/\mathfrak{q}^{\nu+1}E$ is an R-module of finite length. Thus $\Delta\chi_{\mathfrak{q}}(E)$ is a polynomial function of degree $\leq s-1$, and hence $\chi_{\mathfrak{q}}(E)$ is what we claimed it to be.

<u>Remark:</u> If $\mathfrak{q}_1$ and $\mathfrak{q}_2$ are two ideals containing some powers of the maximal ideal $\mathfrak{m}$, i.e. if

$\mathfrak{q}_1 \supset \mathfrak{m}^{n_1}$ and $\mathfrak{q}_2 \supset \mathfrak{m}^{n_2}$, then the degrees of $\chi_{\mathfrak{q}_1}(E)$ and $\chi_{\mathfrak{q}_2}(E)$ are equal. In particular, they are all equal to the degree of $\chi_{\mathfrak{m}}(E)$. This is seen easily by observing that for any integer n, $\chi_{\mathfrak{m}^n}(E)$ and $\chi_{\mathfrak{m}}(E)$ have the same degree. Thus, if $\mathfrak{m}^n < \mathfrak{q} < \mathfrak{m}$, we must have $\deg(\chi_{\mathfrak{m}^n}(E)) \leq \deg(\chi_{\mathfrak{q}}(E)) \leq \deg(\chi_{\mathfrak{m}}(E))$ and hence equality.

Definition. If R is a local ring, the dimension of R is the degree of the polynomial function $\chi_{\mathfrak{m}}(R)$. The dimension of an R-module E is the degree of the polynomial function $\chi_{\mathfrak{m}}(E)$. If R is a noetherian ring and $\mathfrak{p}$ is a prime ideal of R, then the height of $\mathfrak{p}$ is the dimension of $R_{\mathfrak{p}}$.

Proposition 3.5. Let R be a local ring, and let s be the smallest number of elements required to generate an ideal $\mathfrak{q}$ of R which contains some power of the maximal ideal $\mathfrak{m}$. Then $\dim R \leq s$. In particular, $\dim R \leq [\mathfrak{m}/\mathfrak{m}^2 : k]$.

Although we will not be using this fact immediately, it is important to note that for a local ring R, the following integers are equal:

a) the dimension of R;

b) the smallest number of elements required to generate an ideal containing a power of the maximal ideal;

c) the length of the longest chain of prime ideals in R where the length of the chain $\mathfrak{p}_0 \supset \mathfrak{p}_1 \supset \ldots \supset \mathfrak{p}_h$ is defined to be the integer h.

The proof that these three integers are equal is not completely trivial. The reader is referred to Zariski-Samuel, Commutative Algebra, for those details about noetherian rings which are mentioned here but not proved.

For various reasons, it is useful to be able to compare the dimension of a module E over a local ring R with that of $E/(x)E$ if x is an element of R. From our point of view, it is extremely helpful to know what happens to $\dim E/(x)E$ when x is not a zero divisor for E. To handle this problem, we quote the Artin-Rees Theorem:

Theorem 3.6. Let R be a noetherian ring, M a finitely generated R-module, M' a submodule of M, and I an ideal of R. Then there exists an integer $h > 0$ such that for all $n \geq h$ we have $(I^n M) \cap M' = I^{n-h}(I^h M \cap M')$.

Corollary 3.7. If R is a local ring and x is an element of the maximal ideal $\mathfrak{m}$ which is a non-zero divisor for an R-module E, there is an integer $h > 0$ such that for all $\nu \geq h$, $M^{\nu}E:x \subset \mathfrak{m}^{\nu-h}E$ where $\mathfrak{m}^{\nu}E:x = \{e \varepsilon E/xe \varepsilon \mathfrak{m}^{\nu}E\}$.

The proof depends upon the Artin-Rees Theorem and the easy observation that $x(M^{\nu}E;x) = \mathfrak{m}^{\nu}E \cap (x)E$. For then we have $x(\mathfrak{m}^{\nu}E:x) = \mathfrak{m}^{\nu}E \cap (x)E = \mathfrak{m}^{\nu-h}(\mathfrak{m}^{h}E \cap (x)E) = x\mathfrak{m}^{\nu-h}(\mathfrak{m}^{h}E:x) \subset x\mathfrak{m}^{\nu-h}E$ from which we conclude: $\mathfrak{m}^{\nu}E:x \subset \mathfrak{m}^{\nu-h}E$.

<u>Proposition 3.8</u>. Let R be a local ring, E an R-module, and x an element of $\mathfrak{m}$. Then $\dim E/(x)E \geq \dim E - 1$. If x is not a zero divisor for E, we have $\dim E/(x)E \leq \dim E - 1$ and hence $\dim E/(x)E = \dim E - 1$.

<u>Proof</u>. If we let $\bar{E} = E/(x)E$, and $\bar{\mathfrak{m}} = \mathfrak{m}/(x)$, we are interested in the length of $\bar{E}/\bar{\mathfrak{m}}^{\nu}\bar{E}$. But $\bar{E}/\bar{\mathfrak{m}}^{\nu}\bar{E} = E/(\mathfrak{m}^{\nu},x)E$, and we have an exact sequence:

$$0 \longrightarrow (\mathfrak{m}^{\nu},x)E/\mathfrak{m}^{\nu}E \longrightarrow E/\mathfrak{m}^{\nu}E \longrightarrow E/(\mathfrak{m}^{\nu},x)E \longrightarrow 0 .$$

Thus length $E/(\mathfrak{m}^{\nu},x)E = \ell(E/\mathfrak{m}^{\nu}E) - \ell((\mathfrak{m}^{\nu},x)E/\mathfrak{m}^{\nu}E)$. Since $(\mathfrak{m}^{\nu},x)E/\mathfrak{m}^{\nu}E \approx (x)E/\mathfrak{m}^{\nu}E \cap (x)E \approx xE/x(\mathfrak{m}^{\nu}E:x)$ we have length $(\mathfrak{m}^{\nu},x)E/\mathfrak{m}^{\nu}E \leq$ length $E/\mathfrak{m}^{\nu}E:x$, with equality holding if x is not a zero divisor for E. Since $x \varepsilon \mathfrak{m}$, we know that $\mathfrak{m}^{\nu-1}E \subset \mathfrak{m}^{\nu}E:x$ so that length $E/\mathfrak{m}^{\nu}E:x \leq$ length $E/\mathfrak{m}^{\nu-1}E$. Thus length $E/(\mathfrak{m}^{\nu},x)E =$ length $E/\mathfrak{m}^{\nu}E -$ length $(\mathfrak{m}^{\nu},x)E/\mathfrak{m}^{\nu}E \geq$ length $E/\mathfrak{m}^{\nu}E -$ length $E/\mathfrak{m}^{\nu+1}E$ so that $\dim \bar{E} \geq \dim E - 1$. If, however, x is not a zero divisor for E, we have length $(\mathfrak{m}^{\nu},x)E/\mathfrak{m}^{\nu}E) =$ length $(E/\mathfrak{m}^{\nu}E:x)$ and by 3.7, length $(E/\mathfrak{m}^{\nu}E:x) \geq$ length $(E/\mathfrak{m}^{\nu-h}E)$ for suitable fixed h and all $\nu \geq h$. Thus length $E/(\mathfrak{m}^{\nu},x)E =$ length $(E/\mathfrak{m}^{\nu}E) -$ length $(\mathfrak{m}^{\nu},x)E/\mathfrak{m}^{\nu}E \leq$ length $(E/\mathfrak{m}^{\nu}E) -$ length $(E/\mathfrak{m}^{\nu-h}E)$ which immediately implies that $\dim \bar{E} \leq \dim E - 1$.

In the above proof we said we were interested in the length of $\bar{E}/\bar{\mathfrak{m}}^{\nu}\bar{E}$ instead of the length of $\bar{E}/\mathfrak{m}^{\nu}\bar{E}$ to emphasize the fact that $\dim \bar{E}$ as an R-module is the same as $\dim \bar{E}$ as an $\bar{R}$-module where $\bar{R} = R/(x)$. This is mainly to ensure that when $E = R$, $\dim \bar{R}$ as an R-module is seen to be the dimension of the local ring $\bar{R}$. Of course, $\bar{E}/\bar{\mathfrak{m}}^{\nu}\bar{E} \approx \bar{E}/\mathfrak{m}^{\nu}\bar{E}$.

As an immediate consequence of 3.8 we have

<u>Corollary 3.9</u>. Let R be a local ring, E an R-module, and $x_1,\ldots,x_s$ an E-sequence. Then

$$\dim E/(x_1,\ldots,x_s)E = \dim E - s.$$

In particular, $s \leq \dim E$.

§4. Codimension and finitistic global dimension

We will now let R be a noetherian ring, E a finitely generated R-module, and $\mathfrak{a}$ an ideal in such that $E/\mathfrak{a}E \neq 0$. Since $E/\mathfrak{a}E \neq 0$, there is some prime ideal $\mathfrak{p}$ in R such that $E/\mathfrak{a}E \otimes_R R_{\mathfrak{p}} \neq 0$, and this prime ideal $\mathfrak{p}$ obviously contains $\mathfrak{a}$. From Corollary 1.8, we see that if $x_1,\dots,x_s$ is an E-sequence contained in $\mathfrak{a}$, then $x_1,\dots,x_s$ is also an $E \otimes_R R_{\mathfrak{p}}$ - sequence in $\mathfrak{a}R_{\mathfrak{p}}$ and thus, by 3.9, $s \leq \dim E \otimes R_{\mathfrak{p}}$. Hence the number of elements in an E-sequence contained in $\mathfrak{a}$ is bounded, and any E-sequence in $\mathfrak{a}$ may be extended to a maximal E-sequence in $\mathfrak{a}$. We shall prove that any two maximal E-sequences in $\mathfrak{a}$ have the same number of elements but first we must review the notion of associated prime ideals.

Definition. If E is an R-module, a prime ideal $\mathfrak{p}$ is said to be associated to E if there is some monomorphism $R/\mathfrak{p} \longrightarrow E$. The associator of E, denoted by $\mathrm{Ass}(E)$, is the set of all primes associated to E.

Using the fact that R is noetherian, we have

Lemma 4.1. If E is an R-module, then $\mathrm{Ass}(E) = \emptyset$ if and only if $E = 0$.

Proof. Clearly if $E = 0$ we have $\mathrm{Ass}(E) = \emptyset$. To show that $E \neq 0$ implies $\mathrm{Ass}(E) \neq \emptyset$, one consider the set of all ideals $\mathfrak{a}$ such that there is a monomorphism $R/\mathfrak{a} \longrightarrow E$. This set of ideals is not empty, and thus there is a maximal such ideal $\mathfrak{a}_0$. One then proves easily that such an ideal $\mathfrak{a}_0$ is necessarily prime.

Another useful lemma is the following:

Lemma 4.2. Let $0 \longrightarrow E' \longrightarrow E \longrightarrow E''$ be an exact sequence of R-modules. Then $\mathrm{Ass}(E) \subset \mathrm{Ass}(E') \cup \mathrm{Ass}(E'')$.

The proof of this lemma is trivial, and one can now prove

Proposition 4.3. If R is a noetherian ring and E is a finitely generated R-module, then $\mathrm{Ass}(E)$ is a finite set.

Proof. One first proves this for cyclic modules, i.e. modules of the form $R/\mathfrak{a}$ where $\mathfrak{a}$ is an ideal of R. One then assumes that the theorem is true for modules generated by less than n + 1 elements, and assuming that E is generated by $e_1,\dots,e_{n+1}$ one considers the exact sequence $0 \longrightarrow E' \longrightarrow E \longrightarrow E'' \longrightarrow 0$ where E' is generated by $e_1,\dots,e_n$. Applying 4.2 one gets the desired result.

Definition. If R is a noetherian ring and E a finitely generated R-module, we define the height of E to be min height $\mathfrak{p}$ where $\mathfrak{p}$ runs through all primes in Ass(E). If $\mathfrak{a}$ is an ideal of R, we generally abuse notation and call the height of $R/\mathfrak{a}$ the height of the ideal $\mathfrak{a}$.

An important result relating the height of an ideal with the number of generators of the ideal, is the Krull Principal Ideal Theorem. This states that if R is a noetherian ring, and $\mathfrak{a}$ is an ideal generated by elements $x_1,\dots x_r$, then any minimal prime of $\mathrm{Ass}(R/\mathfrak{a})$ has height less than or equal to r. In particular, height $\mathfrak{a} \leq r$. This follows from 3.5 since if $\mathfrak{p}$ is minimal in $\mathrm{Ass}(R/\mathfrak{a})$, then $\mathfrak{a}R_{\mathfrak{p}}$ contains some power of $\mathfrak{p}R_{\mathfrak{p}}$ and thus height $\mathfrak{p} = \dim R_{\mathfrak{p}} \leq r$.

Remarks: 1) For any R-module E, the set of zero divisors of E is the union of all primes in Ass(E).

(2) Thus, if E is a finitely generated R-module (with R noetherian) and if $\mathfrak{a}$ is an ideal such that every element of $\mathfrak{a}$ is a zero divisor of E, then there is a non-zero element e of E such that $\mathfrak{a}e = 0$. For if $\mathfrak{a}$ is as above, then $\mathfrak{a} \subset \cup \mathfrak{p}$ where $\mathfrak{p}$ runs through Ass(E) and hence $\mathfrak{a} \subset \mathfrak{p}$ for some $\mathfrak{p} \in \mathrm{Ass}(E)$ (since Ass(E) is finite). We therefore have the epimorphism $R/\mathfrak{a} \longrightarrow R/\mathfrak{p}$ and a monomorphism $R/\mathfrak{p} \longrightarrow E$. The image of 1 under the composite map $R \longrightarrow R/\mathfrak{a} \longrightarrow R/\mathfrak{p} \longrightarrow E$ is the desired element $e \in E$.

(3) If the ideal $\mathfrak{a}$ in Remark 2 above is a maximal ideal then $\mathfrak{a} \in \mathrm{Ass}(E)$.

We are now ready to prove the main theorem of this section.

Theorem 4.4. Let R be a noetherian ring, E an R-module, and $\mathfrak{a}$ an ideal of R generated by elements $x_1,\dots,x_n$ such that $E/\mathfrak{a}E \neq 0$. Let $y_1,\dots,y_s$ be a maximal E-sequence in $\mathfrak{a}$. Then $s+q = n$, where q is the dimension of the highest non-vanishing homology of the complex $E(x_1,\dots,x_n)$. Furthermore,

$$H_q(E(x_1,\dots,x_n)) \approx (y_1,\dots,y_s)E : \mathfrak{a}/(y_1,\dots,y_s)E \quad .$$

Proof. The proof proceeds by induction on s. When $s=0$, it means that every element of $\mathfrak{a}$ is a zero divisor for E so that by Remark 2 above, there is a non-zero element e in E such that $\mathfrak{a}e = 0$. Since $H_n(E(x_1,\dots,x_n)) = 0:\mathfrak{a} \neq 0$, we see that $q = n$ and $H_n(E(x_1,\dots,x_n)) = (y_1,\dots, y_s)E:\mathfrak{a}/(y_1,\dots, y_s)E = 0:\mathfrak{a}$.

When $s > 0$, we consider the exact sequence $0 \longrightarrow E \xrightarrow{y_1} E \longrightarrow \bar{E} \longrightarrow 0$ and get the

$$H_{\bar{q}+1}(\bar{E}(x_1,\dots,x_n)) \longrightarrow H_{\bar{q}}(E(x_1,\dots,x_n)) \xrightarrow{y_1} H_q(E(x_1,\dots,x_n)) \longrightarrow H_{\bar{q}}(\bar{E}(x_1,\dots,x_n)) \longrightarrow \dots ,$$

where $\bar{q}$ is the dimension of the highest non-vanishing homology group of $\bar{E}(x_1,\dots,x_n)$. Thus, $H_{\bar{q}+1}(\bar{E}(x_1,\dots,x_n)) = 0$, $H_{\bar{q}}(\bar{E}(x_1,\dots,x_n)) \neq 0$, and multiplication by y_1 on $H_{\bar{q}}(E(x_1,\dots,x_n))$ and on $H_{\bar{q}-1}(E(x_1,\dots,x_n)$ is zero since y_1 is in $\mathfrak{a}$. Thus $H_{\bar{q}}(E(x_1,\dots,x_n)) = 0$ while $H_{\bar{q}-1}(E(x_1,\dots,x_n)) \approx H_{\bar{q}}(\bar{E}(x_1,\dots,x_n))$. Noting that $y_2,\dots, y_s$ is a maximal E-sequence in $\mathfrak{a}$, and using induction, we have $(s-1) + \bar{q} = n$. Since, however, we have just shown that $q = \bar{q}-1$, we have $s+q = n$. Finally, since $H_{\bar{q}}(\bar{E}(x_1,\dots,x_n)) \approx (y_2,\dots, y_s)\bar{E}: \mathfrak{a}/(y_2,\dots, y_s)\bar{E}$ $\approx (y_1,\dots, y_s)E: \quad /(y_1,\dots, y_s)E$, and since $H_q(E(x_1,\dots,x_n)) = H_{\bar{q}-1}(E(x_1,\dots,x_n)) \approx H_{\bar{q}}(\bar{E}(x_1,\dots,x_n$ we have

$$H_q(E(x_1,\dots,x_n)) \approx (y_1,\dots, y_s)E: \mathfrak{a}/(y_1,\dots, y_s)E.$$

Corollary 4.5. If R, E, and $\mathfrak{a}$ are as in 4.4, then any two maximal E-sequences in $\mathfrak{a}$ have the same length.

Definition. The length of a maximal E-sequence in $\mathfrak{a}$ is called the $\mathfrak{a}$-depth of E, denoted by depth $(\mathfrak{a}:E)$. If R is a local ring, then codim E = depth $(\mathfrak{m}:E)$ where $\mathfrak{m}$ is the maximal ideal of R.

An important consequence of 4.4 is

Theorem 4.6. Let R be a local ring, and E an R-module such that $\mathrm{hd}_R E < \infty$. Then

$$\operatorname{codim} R = \mathrm{hd}_R E + \operatorname{codim} E \quad .$$

Proof. Let $\mathfrak{m} = (x_1,\dots,x_n)$, where $\mathfrak{m}$ is the maximal ideal of R. Denote by q the dimension of the highest non-vanishing homology group of $K(x_1,\dots,x_n)$, and by q_E the corresponding integer for the complex $E(x_1,\ldots,x_n)$. Since codim R = n-q and codim E = $n-q_E$, we want to show that $q_E - q = hd_R E$.

Now when $hd_R E = 0$, E is free and clearly $q_E = q$. Suppose that $hd_R E \geq 1$. Then we have an exact sequence $0 \longrightarrow L \longrightarrow F \longrightarrow E \longrightarrow 0$ with F a free module, and $hd_R E = 1 + hd_R L$. This gives us an exact sequence:

$$H_{1+q_L}(E(x_1,\ldots,x_n)) \longrightarrow H_{q_L}(L(x_1,\ldots,x_n)) \longrightarrow H_{q_L}(F(x_1,\ldots,x_n)) \longrightarrow H_{q_L}(E(x_1,\ldots,x_n))$$

$$\longrightarrow H_{q_L-1}(L(x_1,\ldots,x_n)).$$

By induction on $hd_R E$, we know that $q_L - q = hd_R L$ so what must be shown is that $1 + q_L = q_E$. If $q_L - q > 0$, then $H_{q_L}(F(x_1,\ldots,x_n)) = 0$ and thus $H_{1+q_L}(E(x_1,\ldots,x_n)) \neq 0$, while clearly $H_{t+q_L}(E(x_1,\ldots,x_n)) = 0$ for all $t > 1$. Thus in this case, we would have $q_E = 1 + q_L$ and we'd be done. Our problem, then, is to resolve the case when $q_L = q$, i.e. when $hd_R L = 0$ or L is free. In this case, we may assume that L and F have been chosen so that $L \subset \mathfrak{m}F$.

If $y_1,\ldots, y_s$ is a maximal R-sequence, it is also a maximal F- and L-sequence, and 4.4 tells us that

$$H_q(L(x_1,\ldots,x_n)) = (y_1,\ldots, y_s)L:\mathfrak{m}/(y_1,\ldots, y_s)L$$

$$H_q(F(x_1,\ldots,x_n)) = (y_1,\ldots, y_s)F:\mathfrak{m}/(y_1,\ldots, y_s)F$$

and the map $H_q(L(x_1,\ldots,x_n)) \longrightarrow H_q(F(x_1,\ldots,x_n))$ is the natural map of $(y_1,\ldots, y_s)L:\mathfrak{m}/(y_1,\ldots, y_s)L \longrightarrow (y_1,\ldots, y_s)F:\mathfrak{m}/(y_1,\ldots, y_s)F$. If we show that this map is not a monomorphism, we are done.

Since $(y_1,\ldots, y_s):\mathfrak{m} \supsetneq (y_1,\ldots, y_s)$, there is a $z \notin (y_1,\ldots, y_s)$ with $z\mathfrak{m} \subset (y_1,\ldots, y_s)$. Since L is free, $zL \not\subset (y_1,\ldots, y_s)L$. But $zL \subset z\mathfrak{m}F \subset (y_1,\ldots, y_s)F$. Choosing an element $w \in zL - (y_1,\ldots, y_s)L$, we have $w \in (y_1,\ldots, y_s)L:\mathfrak{m}$ but $w \notin (y_1,\ldots, y_s)L$ and thus $\overline{w} \neq 0$ in $(y_1,\ldots, y_s)L:\mathfrak{m}/(y_1,\ldots, y_s)L$. However, $\overline{w}$ is mapped to zero in $(y_1,\ldots, y_s)F:\mathfrak{m}/(y_1,\ldots, y_s)F$ so that our map $H_q(L(x_1,\ldots,x_n)) \longrightarrow H_q(F(x_1,\ldots,x_n))$ is not a monomorphism, and the proof is complete.

A particular consequence of 4.6 is that if R is a local ring, and E an R-module such that $hd_R E < \infty$, then $hd_R E \leq \text{codim } R$.

Definition: If R is a commutative ring, we define the finitistic global dimension of R, written f.gl.dim R, to be $\sup hd_R E$ where E ranges over all finitely generated R-modules of finite homological dimension.

Before winding up this section, we state the following technical lemma:

Lemma 4.7. If R is a local ring, E a finitely generated R-module, and $x \in \mathfrak{m}$ an element which is not a zero divisor for E, then $hd_R E/xE = 1 + hd_R E$.

Proposition 4.8. If R is a local ring, then f.gl.dim R = codim R.

Proof. We have already seen that f.gl.dim $R \leq$ codim R. However, if $y_1, \ldots, y_s$ is an R-sequence, with s = codim R, we have (by 4.7) $hd_R R/(y_1, \ldots, y_s) = s$ = codim R and hence the equality.

Putting 2.6, 3.5, 3.9, and 4.8 together, we obtain

Theorem 4.9. If R is a local ring, we have

$$\text{f.gl.dim } R = \text{codim } R \leq \dim R \leq [\mathfrak{m}/\mathfrak{m}^2 : k] \leq \text{gl.dim } R.$$

§5. Regular local rings.

We are now ready to apply what we have done to the study of s me important properties of regular local rings.

Definition. A local ring R is regular if $\dim R = [\mathfrak{m}/\mathfrak{m}^2 : k]$.

The following is a well-known property of regular local rings.

Proposition 5.1. Let R be a regular local ring, and $x_1, \ldots, x_n$ a minimal generating set of the maximal ideal $\mathfrak{m}$. Then for each integer L, the ideal $(x_1, \ldots, x_i)$ is a prime ideal. In particular, $x_1, \ldots, x_n$ is a (maximal) R-sequence.

Our first main result about regular local rings is:

Theorem 5.2. A local ring R is regular if and only if gl.dim $R < \infty$. If R is regular, gl.dim R = dim R.

Proof. If R is regular, 5.1 tells us that $hd_R k = hd_R R/(x_1,\ldots,x_n) = n$ = dim R, and hence gl.dim $R = n < \infty$. Conversely, if gl.dim $R < \infty$, then gl.dim R = f.gl.dim R and by 4.9, we have dim $R = [\mathfrak{m}/\mathfrak{m}^2 : k]$.

Using this characterization of regular local rings, we obtain

Theorem 5.3. If R is a regular local ring and $\mathfrak{p}$ is a prime ideal of R, then $R_\mathfrak{p}$ is a regular local ring.

Proof. gl.dim $R_\mathfrak{p} = hd_{R_\mathfrak{p}} R_\mathfrak{p}/\mathfrak{p}R_\mathfrak{p} \le hd_R R/\mathfrak{p} < \infty$. Thus R is regular.

Lemma 5.4. If R is a regular local ring, E a finitely generated R-module, and $\mathfrak{p} \in$ Ass(E), then $hd_R E \geq$ height $\mathfrak{p}$.

Proof. We have $hd_R E \ge hd_{R_\mathfrak{p}} E_\mathfrak{p}$. Since $\mathfrak{p}R_\mathfrak{p}$ is obviously in Ass(E), and $\mathfrak{p}R_\mathfrak{p}$ is the maximal ideal of $R_\mathfrak{p}$, we have codim E = 0 as an $R_\mathfrak{p}$-module. Thus $hd_{R_\mathfrak{p}} E_\mathfrak{p}$ = codim $R_\mathfrak{p}$ = dim $R_\mathfrak{p}$ = height $\mathfrak{p}$.

A straightforward application of the Krull Principal Ideal Theorem yields the following

Lemma 5.5. If R is a noetherian ring, $\mathfrak{a}$ an ideal of R, and x an element of R such that x is not a zero divisor for $R/\mathfrak{a}$ and $R/(\mathfrak{a},x) \neq 0$, then height $(\mathfrak{a},x) \geq 1 +$ height $\mathfrak{a}$. In particular, if $x_1,\ldots,x_s$ is an R-sequence, then height $(x_1,\ldots,x_s) = s$.

Lemma 5.6. Let   be an ideal in a regular local ring R. Then depth (;R) = height .

Proof. By localizing we know that depth $(\mathfrak{a};R) \leq$ height $\mathfrak{a}$. Suppose height $\mathfrak{a} = s$ and let $x_1,\ldots,x_t$ be a maximal R-sequence in $\mathfrak{a}$ i.e. t = depth $(\mathfrak{a};R)$. Since every element of $\mathfrak{a}$ is a zero-divisor for $R/(x_1,\ldots,x_t)$, we must have $\mathfrak{a}$ contained in some prime ideal $\mathfrak{p} \in$ Ass$(R/(x_1,\ldots,x_t))$. But if $\mathfrak{p} \in$ Ass$(R/(x_1,\ldots,x_t))$ we have, by 5.4, height $\mathfrak{p} \leq hd_R R/(x_1,\ldots,x_t) = t$. Thus, since $\mathfrak{a} \subset \mathfrak{p}$ implies height $\mathfrak{a} \leq$ height $\mathfrak{p}$, we have $s \leq t$ and thus $s = t$.

Lemma 5.7. Let R be a noetherian ring, E a finitely generated R-module, and x a non-zero divisor for E. If $\mathfrak{p}$ is in Ass(E) and $\mathfrak{p}'$ is a prime ideal containing $(\mathfrak{p},x)$, then there is a prime ideal $\mathfrak{p}'' \in$ Ass(E/xE) such that $\mathfrak{p}' \supset \mathfrak{p}'' \supset \mathfrak{p}$.

Definition. We say a chain of prime ideals $\mathfrak{p}_0 \subset \mathfrak{p}_1 \subset \dots \subset \mathfrak{p}_n$ is saturated if for each i there is no prime lying strictly between $\mathfrak{p}_i$ and $\mathfrak{p}_{i+1}$. We say a ring satisfies the saturated chain condition for prime ideals (s.c.c.) if any two saturated chains of primes between any two given primes $\mathfrak{p} \subset \mathfrak{p}'$ have the same length.

Theorem 5.8. Every regular local ring satisfies the saturated chain condition for prime ideals.

Proof. It obviously suffices to show that if $\mathfrak{p} \subset \mathfrak{p}'$ is saturated, then height $\mathfrak{p}' = 1 +$ height . Suppose, then, that s = height $\mathfrak{p}$. Then there is an R-sequence $x_1,\dots,x_s$ which is maximal in $\mathfrak{p}$. Clearly there is an element x_{s+1} in $\mathfrak{p}'-\mathfrak{p}$ such that $x_1,\dots,x_{s+1}$ is an R-sequence. Using 5.7, there is a prime $\mathfrak{p}'' \in \mathrm{Ass}(R/(x_1,\dots,x_{s+1}))$ such that $\mathfrak{p}' \supset \mathfrak{p}'' \supset \mathfrak{p}$. Since height " = s+1 > height we cannot have $\mathfrak{p}'' = \mathfrak{p}$. Thus $\mathfrak{p}'' = \mathfrak{p}'$ and height $\mathfrak{p}' = s+1 = 1 +$ height $\mathfrak{p}$.

Remarks: 1) By constructing a local ring that does not satisfy the s.c.c., Nagata was able to show that not every local ring is a factor ring of a regular local ring. For clearly, a factor ring of a ring satisfying the s.c.c. must also satisfy the s.c.c.

2) We see that if R is a regular local ring, then dim $R/\mathfrak{p}$ + height $\mathfrak{p}$ = n = dim R.

3) It can be shown that if R satisfies the s.c.c. and if $\mathfrak{a}$ is an ideal of R and x and element of R, then height $(\mathfrak{a},x) \leq 1 +$ height $\mathfrak{a}$.

We now come to the Cohen-Macaulay Theorem:

Theorem 5.9. Let R be a regular local ring, and $\mathfrak{a} = (x_1,\dots,x_s)$ an ideal of height s. Then $\mathfrak{a}$ is unmixed (i.e. every prime $\mathfrak{p} \in$ Ass $(R/\mathfrak{a})$ has height s), and $x_1,\dots,x_s$ is an R-sequence.

Proof. The main point of the proof is to show that $x_1,\dots,x_s$ is an R-sequence, for then the rest follows easily. To show that $x_1,\dots,x_s$ is an R-sequence, we proceed by induction. Certainly when s=1 we are done (since a regular local ring is an integral domain). Assuming s > 1, we have s = height $(x_1,\dots,x_s) \leq 1 +$ height $(x_1,\dots,x_{s-1})$. But then $s-1 \leq$ height $(x_1,\dots,x_{s-1}) \leq s-1$ so by induction we have $x_1, \dots, x_{s-1}$ is an R-sequence. It is then trivial to show that x_s is not a zero divisor for $R/(x_1, \dots, x_{s-1})$ (by using a height argument), and so we have the result.

We end these lectures with an outline of a proof of the fact that every regular local ring is a unique factorization domain. The proof given here is due to Kaplansky and is extremely elegant. As the reader will see, the tone of the proof is very different from most of what has gone before. It would perhaps be nice if a proof could be found using the notion of R-sequences.

Definitions. An element x of an integral domain R is a prime element if the ideal (x) is a prime ideal. An integral domain is a unique factorization domain (ufd), if every non-unit of R is a finite product of prime elements of R.

Theorem 5.10. Every regular local ring R is a unique factorization domain.

Outline of proof. We have already noted that a regular local ring is an integral domain. Also, R contains prime elements since every element which is part of a minimal generating set for the maximal ideal of R is prime. Let x be such a prime element and let S be the multiplicative set consisting of all non-negative powers of x. Then it can be shown that R is a UFD if and only if R_S is a UFD. Hence we must show that R_S is a UFD. Since R_S is a noetherian domain, it suffices to show that every minimal prime ideal of R_S (i.e. every prime ideal of height 1) is principal. If we could show that every such ideal were free, it would follow that it is principal. As a first attempt at showing freeness, we show that minimal prime ideals are projective. Let $\mathcal{Y}$ be such an ideal, and $\mathfrak{m}$ any maximal ideal of $R' = R_S$. Then since $\text{gl.dim. } R'_{\mathfrak{n}} \leq \text{gl.dim. } R' \leq \text{gl.dim. } R < \infty$, we have that R' is a regular local ring. Furthermore, it is easy to see that $\dim R'_{\mathfrak{n}} < \dim R$, so that by induction on $\dim R$, we know that $R'_{\mathfrak{n}}$ is a UFD. Hence $\mathcal{Y}R'_{\mathfrak{n}}$, being either a minimal prime ideal of $R'_{\mathfrak{n}}$ or $R'_{\mathfrak{n}}$ itself, is free. Since $\text{hd}_{R'}\mathcal{Y} = \sup \text{hd}_{R'_{\mathfrak{n}}}\mathcal{Y}R'_{\mathfrak{n}}$ when $\mathfrak{n}$ ranges over all maximal ideals of R', we see that $\text{hd}_{R'}\mathcal{Y} = 0$ and $\mathcal{Y}$ is projective.

Next we show that projective ideals of R' are principal. To see this, we first observe that if E' is an R'-module, then E' has a finite free resolution. For $E' = E \otimes R'$ for some R-module E, and choosing a free resolution for E over R, we obtain one for E' by tensoring with R'. In particular, it is easy to see now that if E is a projective R'-module, then E has a free resolution of length one i.e. we can find free R'-modules F_0 and F_1 such that $E \oplus F_1 = F_0$. Finally, when E is a projective ideal it is easy to see that $\overset{p}{\Lambda} E = 0$ for $p > 1$ since $(\overset{p}{\Lambda} E)_{\mathfrak{n}} \approx \overset{p}{\Lambda} E_{\mathfrak{n}} = 0$ for $p > 1$ and all maximal ideals $\mathfrak{n}$ of R' (because $E_{\mathfrak{n}}$ is principal for all such $\mathfrak{n}$). Applying the fact that $\Lambda^{n}(A \oplus B) \approx \Sigma\Lambda^{p} A \otimes \Lambda^{n-p} B$ to the situation $E \oplus F_1 = F_0$, and letting n = rank of F_0, we have $R' \approx \Lambda^{n} F_0 \approx \Lambda^{1} E \otimes \Lambda^{n-1} F_1 \approx \Lambda^{1} E \otimes R'$ (since F_1 must have rank n-1), and thus $R' \approx \Lambda^{1} E = E$. Since all projective ideals of R' are principal, and all minimal prime ideals of R' are projective, we have shown that R' is a UFD. Thus R, too, is a UFD, and the outline of the proof is complete.

Since all the results mentioned in these notes are known, I have not given a detailed bibliography Most of what has been said has its source in papers by Serre and Auslander-Buchsbaum with background material and some amplification in the already mentioned volume by Zariski-Samuel. Perhaps the materia on polynomial functions is new and not yet in print. It will appear in a paper by Marshall Fraser concerned with multiplicities, but the elements and purely formal properties of generalized polynomial functions have also been discussed by P. Wagreich.

CATEGORIES OF MODELS WITH INITIAL OBJECTS

by

Dr. R. Fittler

INTRODUCTION

The construction of free algebras with finitary operations in the sense of Birkhoff (cf. [1], P.M. Cohn) can be viewed as a prototype for the construction of more general universal structures. In order to motivate the subsequent concepts we need a little modification of the definition of free algebras. Let T be a fixed algebraic theory . Usually an algebra φ is called freely generated by a set N if there is a map $i: N \longrightarrow |\varphi|$ (underlying set of φ) such that for each map $f: N \longrightarrow |M|$, where M is an arbitrary algebra of type T there is a uniquely determined homomorphism $g: \varphi \longrightarrow M$ which makes the diagram

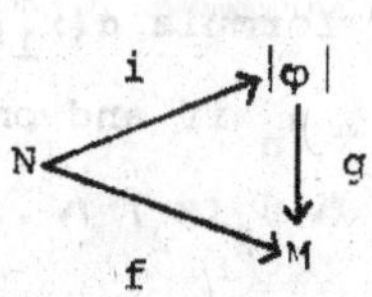

commutative. The model M of T together with $f: N \longrightarrow |M|$ can be viewed as a model M' of the (algbraic) theory T' consisting of the theory T together with new individual constants, one for each element of N (see definition 14). The model φ' (φ together with $i: N \longrightarrow |\varphi|$) has the property that for each model M' of T' there is exactly one homomorphism $g: \varphi' \longrightarrow M'$ (in the sense of the algebraic theory T'). This suggests to define an initial model φ of a first order theory T by the condition that for each model M of T there is exactly one map $g: \varphi \longrightarrow M$, which fulfills certain conditions. More precisely these maps are maps $|\varphi| \longrightarrow |M|$ between the underlying sets which preserve a given set F of formulas of T (see 2). They are called F-maps. φ will be called F-initial.

The main theorem of this paper (characterization theorem 9) states that one can construct the (essentially unique) F-initial model φ , if it exists, almost explicitly in the following way. The underlying set of φ is a quotient set of the set of all unary formulas μ(see 4) each having the property that one can prove in T the sentence
$\exists x \; \mu(x) \wedge \forall x \forall y \; (\mu(x) \wedge \mu(y) \Longrightarrow x = y)$ (i.e., there is exactly one element which fulfills $\mu(x)$). Two such formulas $\mu(x)$, $\nu(x)$ are identified if $\forall x(\mu(x) \Longleftrightarrow \nu(x))$ is provable in T. Furthermore, an n-ary formula $\alpha(x_1, \ldots, x_n)$ of F holds in φ for the n-tuple $\mu_1, \ldots, \mu_n$ if and only if the formula

$$\forall x_1 \; \forall x_2 \ldots \forall x_n \; (\mu_1(x_1) \wedge \mu_2(x_2) \wedge \ldots \wedge \mu(x_n) \Longrightarrow \alpha(x_1, \ldots, x_n))$$

is provable in T.

A model φ which is constructed in this way is called a F-generated model of T. (see definition 7). Examples are tensor product of modules, localization of rings, direct limits in the category of all models of a first order theory T where the morphisms are F-maps.

Another class of examples is obtained by evaluating at left adjoints of appropriate functors between categories of all models of two first order theories T, T'. The functors in consideration result from interpreting the theory T in the theory T' by means of a map $t: T \longrightarrow T'$ of the theories (see definition 10 and remark 12). If T and T' are algebraic theories in the sense of Birkhoff, one gets algebraic functors in the sense of Lawvere [3]. We will use subsequently Gödel's completeness theorem (cf. [2]) without mentioning it.

CATEGORIES OF MODELS

1. Let S, T be first order theories with equality. Let p, q, r, ... denote their predicate constants and a, b, c, ... the individual constants. By α,β,μ,ν we denote formulas and by x, y individual variables.

2. Definition of Generalized Formulas

A set F of formulas of a theory T is called a generalized formula if it contains the formulas x = y for each pair x, y of individual variables and if it is closed:

(i) Under substitution of individual variables by individual variables and individual constants.

(ii) Under conjunction

(iii) Under disjunction

(iv) Under existential quantification.

3. Remark

A generalized formula is a special case of what H.J. Keisler calls a "generalized atomic set of formulas" (cf. [3]).

4. Let us introduce the category $\mathfrak{M}_{F,T}$. Its objects are models M_T, N_T, ... of T and its morphisms $f: M_T \longrightarrow N_T$ are the maps between the underlying sets $|M_T| \longrightarrow |N_T|$ which preserve all formulas of F, i.e., if for $\alpha \in F$ the formula $\alpha(m_1, \ldots, m_n)$ holds in M_T where $m_i \in |M_T|$ then $\alpha(f(m_1, \ldots, f(m_n))$ holds in N_T. In [3] Keisler calls these maps F-expansions. We call them F-maps. The isomorphisms of $\mathfrak{M}_{F,T}$ are called F-isomorphisms. By $F[M_T, N_T]$ we denote the set of F-maps $M_T \longrightarrow N_T$.

5. Definition of F-initial Models

Let F be a generalized formula of the theory T. A model φ of T is called F-initial if for each model M_T of T there is one unique F-map $g: \varphi \longrightarrow M_T$.

6. Remark

All F-initial models of T are F-isomorphic.

7. <u>Definition of F-generated Models</u>

Let T be a theory and F a generalized formula of T. We try to construct a model of T in the following way. Let ψ be the set of unary formulas $\mu \in F$ with the property that $\exists x \; \{\mu(x) \wedge \forall y(\mu(y) \Longrightarrow y = x)\}$ is provable in T. Let us call them determinant formulas. We define an equivalence relation on ψ by setting $\mu \approx \gamma$ if the sentence $\forall x \forall y(\mu(x) \wedge \nu(y) \Longrightarrow x = y)$ is provable in T. Let $\phi = \{(\mu), (\nu), \ldots\}$ be the set of equivalence classes of ψ. For $\alpha(x_1, \ldots, x_n) \in F$ we define "$\alpha((\mu_1), \ldots, (\mu_n))$ holds in ϕ" if the sentence

$$\forall x_1 \; \forall x_2 \ldots \; \forall x_n \; (\mu_1(x_1) \wedge \ldots \wedge \mu_n(x_n) \Longrightarrow \alpha(x_1, \ldots, x_n))$$

is provable in T.

In general this definition is not consistent. If it is consistent and if there is an interpretation of the remaining predicate constants of T in ϕ such that one obtains a model φ of T then φ is called an <u>F-generated model</u> of T.

8. <u>Remark</u>

All F-generated models of T are F-isomorphic.

9. <u>Characterization Theorem</u>

A model φ of T is an F-generated model of T if and only if φ is F-initial.

Proof. First let φ be F-generated. The elements of $|\varphi|$ being equivalence classes (μ) of determinant formulas μ, we define $g((\mu))$ as the only element $m \in M_T$ such that $\mu(m)$ holds in M_T. Thus g is well defined. Because of the structure of φ it is clear that g is a F-map and it is the only one.

In order to show that a F-initial model of T is F-generated it is necessary to find a determinant formula μ_a for each element $a \in |\varphi|$. For this purpose let S be the following theory. Its language contains the language of T and two unary predicate constants p_a and q_a for each element $a \in |\varphi|$. The axioms of S include those of T and for each n-tuple $a_1, \ldots, a_n$ of elements of $|\varphi|$ and each $\alpha(x_1, \ldots, x_n) \in F$ the two sentences:

$$\forall x_1 \forall x_2 \ldots \forall x_n (p_{a_1}(x_1) \wedge \ldots \wedge p_{a_n}(x_n) \Longrightarrow \alpha(x_1, \ldots x_n))$$

$$\forall x_1 \ldots \ldots \forall x_n (q_{a_1}(x_1) \wedge \ldots \wedge q_{a_n}(x_n) \Longrightarrow \alpha(x_1, \ldots x_n))$$

if $\alpha(a_1, \ldots, a_n)$ holds in φ, and furthermore the sentences

$$\exists x\, p_a(x) \wedge \exists x\, q_a(x) \quad \text{for each } a \in |\varphi|.$$

S is consistent because φ is a model, p_a and q_a both being interpreted by the singleton a, for each $a \in |\varphi|$.

It follows that the sentence $(\forall x)\ (p_a(x) \Longleftrightarrow q_a(x))$ is provable in S for each $a \in |\varphi|$. Because if it were not there would exist a model N of S with different interpretations for some p_a and q_a. Hence there would exist two different maps f_1 and $f_2 : \varphi \longrightarrow N$ assigning to each $a \in |\varphi|$ an element $f_1(a)$ of p_a in N and an $f_2(a)$ of q_a in N respectively. Viewing N as a model of T it follows that f_1 and f_2 are F-maps, in contradiction to the assumption that φ is F-initial.

For a fixed element $a_1 \in |\varphi|$ the proof of the sentence $\forall x(p_{a_1}(x) \Longleftrightarrow q_{a_1}(x))$ in S uses only finitely many axioms. Those which are not already in T are contained in a finite collection C of sentences like:

$$\forall x_1 \ldots \forall x_k (p_{a_1}(x_1) \wedge \ldots \wedge p_{a_k}(x_k) \Longrightarrow \beta(x_1, \ldots, x_k))$$

$$\forall x_1 \ldots \forall x_k (q_{a_1}(x_1) \wedge \ldots \wedge q_{a_k}(x_k) \Longrightarrow \beta(x_1, \ldots, x_k))$$

$\exists x\, p_{a_1}(x), \ldots, \exists x p_{a_k}(x)$ and

$\exists x\, q_{a_1}(x), \ldots, \exists x q_{a_k}(x)$.

where β is an appropriate conjunction of formulas α from F.

Let the theory $\bar{S}$ consist of T together with the collection C of axioms. We know already that $\forall x(p_{a_1}(x) \Longleftrightarrow q_{a_1}(x))$ is provable in $\bar{S}$.

We claim that the unary formula $\mu(x_1)$, defined by

$$\exists x_2 \ \exists x_3 \ldots \exists x_k \quad \beta(x_1, x_2, \ldots, x_k)$$

is a determinant formula in T. It is clear that the formula μ is contained in F and that $\mu(a_1)$ holds in φ. The formula $\exists x\, \mu(x)$ is provable in T because it is true in φ, hence in every model. It remains to show that $\forall x \forall y(\mu(x) \wedge \mu(y) \Longrightarrow x = y)$ is provable in T. Assume it were not; then there would exist a model N of T and two different elements $b_1, c_1 \in |N|$ such that $\mu(b_1)$ and $\mu(c_1)$ hold in N, i.e., there are elements $b_2, \ldots, b_k$ and $c_2, \ldots, c_k$ in N such that $\beta(b_1, b_2, \ldots, b_k)$ and $\beta(c_1, c_2, \ldots, c_k)$ hold in N. Thus N could be extended to a model $\overline{N}$ of $\overline{S}$ by interpreting

$$p_{a_1}(x) \text{ by } x = b_1,\ p_{a_2}(x) \text{ by } x = b_2,\ \ldots,\ p_{a_k}(x) \text{ by } x = b_k$$

$$q_{a_1}(x) \text{ by } x = c_1,\ q_{a_2}(x) \text{ by } x = c_2,\ \ldots,\ q_{a_k}(x) \text{ by } x = c_k$$

which violates $\forall x(p_{a_1}(x) \Longleftrightarrow q_{a_1}(x))$.

We get all determinant formulas in this way because φ is a model of T.

Because φ is F-initial a formula $\alpha(a_1, \ldots, a_n)$ holds in φ (for $\alpha(x_1, \ldots, x_n) \in F$ and $a_1, \ldots, a_n \in |\varphi|$) if and only if the formula:

$$\forall x_1, \ldots, \forall x_n(\mu_{a_1}(x_1) \wedge \ldots \wedge \mu_{a_n}(x_n) \Longrightarrow \alpha(x_1, \ldots, x_n))$$

holds in each model of T, i.e., if and only if it is provable in T.

APPLICATIONS

10. **Definition** of a map $t: T \longrightarrow T'$ of a theory T into a theory T'

t assigns to each n-ary formula α of T a n-ary formula $t(\alpha)$ of T' subject to the following conditions:

(10.1) α and $t(\alpha)$ have the same free variables.

(10.2) t preserves $x = y$ and $x = a$.

(10.3) t preserves the structure of the formulas (e.g., $t(\neg\alpha) = \neg t(\alpha)$, $t[\forall x\alpha\ (x)] = \forall x t(\alpha(x))$, etc.).

(10.4) If the sentence α is provable in T then $t(\alpha)$ is provable in T'.

For the sake of simplicity we assume that T' contains the individual constants and variables of T'.

11.

A map $t: T \longrightarrow T'$ between theories induces in an obvious way an interpretation of the models $M_{T'}$ of T' as models of T.

12. Remark

If F and F' are generalized formulas of T and T' respectively, and if, moreover, $t(F) \subseteq F'$ then the map $t: T \longrightarrow T'$ induces a functor $t^*: \mathfrak{M}_{F',T'} \longrightarrow \mathfrak{M}_{F,T}$ commuting with the underlying set functors.

13.

For a language T containing equality but without additional axioms and a generalized formula F, Keisler observes in [3] that the F-maps $M_T \longrightarrow N_T$ with M_T fixed are in one-one correspondence with the models of a certain theory, say $S_M(F,T)$. The predicate constants of the theory $S_M(F,T)$ are those of T. There is one individual constant for each element $m \in |M_T|$. The axioms are the sentences holding in M_T which are constructed by replacing the free variables of formulas $\alpha \in F$ by some elements of M_T. In order to generalize that, let $t^*: \mathfrak{M}_{F',T'} \longrightarrow \mathfrak{M}_{F,T}$ be a functor defined as in 12.

14. Definition of $S_M(F,T')$

Let M_T be a model of T. The theory $S_M(F,T')$ has the same predicate constants as T'. The individual constants of $S_M(F,T')$ are those of T' and additionally there is a new one for each element $m \in |M_T|$. We do not distinguish between these elements and the corresponding individual constants.

the axioms of $S_M(F,T')$ include those of T' and the sentences of the form $(t\alpha)(m_1, \ldots, m_n)$ where $\alpha(x_1, \ldots, x_m) \in F$, the m_i are elements of M_T and $\alpha(m_1, \ldots, m_n)$ holds in M_T.

15. <u>Lemma</u>

The pairs $(f, N_{T'})$, where $f: M_T \longrightarrow t^*(N_{T'})$ is an F-map are in one-one correspondence with the models of $S_M(F,T')$.

<u>Proof.</u> Let us just establish the correspondence in both directions. A pair $(f: M_T \longrightarrow t^*(N_{T'}), N_{T'})$ is a description of $N_{T'}$ as a model of $S_M(F,T')$, the interpretation of the new individual constants $m \in |M_T|$ given by the map f.

Conversely, if L is a model of $S_M(F,T')$, we get a model $L_{T'}$ of T' by forgetting the interpretation of the individual constants $m \in |M_T|$ (T'-reduct). The map $f: M_T \longrightarrow t^*L_{T'}$ is given by the interpretation of the individual constants $m \in |M_T|$ in L.

Q.E.D.

Let $t: T \longrightarrow T'$ be a map of theories and let F and F' be generalized formulas in T and T' respectively such that $t(F) \subseteq F'$. By Remark 12, this gives rise to a functor:

$$t^*: \mathfrak{M}_{F',T'} \longrightarrow \mathfrak{M}_{F,T} \quad .$$

16. Definition of t*-Free Models Generated by M_T

A model $\varphi \in \mathfrak{M}_{F',T'}$ is called a t*-free model generated by M_T if there is an F-map $i: M_T \longrightarrow t^*\varphi$ such that the correspondence

$$j: F'[\varphi, N_{T'}] \longrightarrow F[M_{T'}, t^*(N_{T'})], \; j(f) = t^*(f) \cdot i$$

is a bijection.

17. Remark

This is the usual universal definition of free objects. It is well known that φ is determined by M_T up to an F'-isomorphism.

If for every $M_T \in \mathfrak{M}_{F,T}$ there is a t*-free model $\varphi(M_T)$ generated by M_T, then φ can be viewed as the left adjoint functor of t* (cf. [5], S. MacLane).

Let $M_T \in \mathfrak{M}_{F,T}$ be a fixed model of T. The generalized formula F' of T' generates a generalized formula F'' of the theory $S_M(F,T')$ by "closing" F' according to the definition 2 of a generalized formula.

18. Lemma

A model φ of T' is a t*-free model generated by $M_T \in \mathfrak{M}_{F,T}$ if and only if the pair $(i: M_T \longrightarrow t^*\varphi, \varphi)$ corresponds (according to Lemma 15) to an F''-initial model of $S_M(F,T')$.

This follows directly from the definitions 5 and 16.

19. Corollary

A model φ of T' is t*-free generated by $M_T \in \mathfrak{M}_{F,T}$ if and only if φ is an F''-generated model of the theory $S_M(F,T')$.

Proof. φ is t*-free generated by M_T if it is F''-initial (by Lemma 17). Because of the characterization theorem 14, we know that φ is an F''-initial model of $S_M(F,T')$ if and only if it is an F''-generated model.

Q.E.D.

REFERENCES

[1] Cohn, P.M. *Universal Algebra*, Harper and Row, New York, 1965.

[2] Gödel, K., Die Vollständigkeit der Axiome des Logischen Funktionen-Kalküls. *Monatsh. Math. Phys.* 37; 349-360, (1930).

[3] Keisler, H.J., "Theory of Models With Generalized Atomic Formulas", *J. of Symbolic Logic*, 25, (1); 1-26, (1960).

[4] Lawvere, F. W., "Functorial Semantics of Algebraic Theories", *Proc. Nat. Acad. Sci.*, 50, (5); 869-872, (1963).

[5] MacLane, S., "Categorical Algebra", *Bull. Amer. Math. Soc.*, 71, (1); 40-106, (1965).

COREFLECTION MAPS WHICH RESEMBLE UNIVERSAL COVERINGS

by

J. F. Kennison

1. INTRODUCTION

If X is connected and semilocally 1-connected (see [9, p. 78]) then one can regard the set $\pi_1(X)$ as any fibre of $c: X^* \longrightarrow X$, where c is the universal simply connected covering of X. This observation generalizes to some non-semilocally 1-connected spaces such as $X = R \times R \setminus \{(2^{-n},0) | n \in Z^+\}$, in the sense that there exists a Hurewicz fibration $c: X^* \longrightarrow X$ (which is a category theory generalization of the universal simply connected covering) such that any fibre of c can be regarded as $\pi_1(X)$. In this last example, moreover, the fibres have an interesting non-trivial topology (see 5.8 the final example of this paper).

Categorically, each of the above maps $c: X^* \longrightarrow X$, can be regarded as a coreflection of X into the category of simply connected and locally pathwise connected spaces. (That is, X^* is simply connected and locally pathwise connected, and if S is any other such space, then every map $S \longrightarrow X$ has a unique lift $S \longrightarrow X^*$ which commutes with c. It is necessary to add that all spaces have base points which are preserved by maps.)

We shall investigate this phenomenon by considering a particular coreflective subcategory $\underline{Cn}^*$ of the category of pointed spaces (i.e. spaces with base points). $\underline{Cn}^*$ is the smallest full replete coreflective subcategory containing all the contractible spaces. (A subcategory is <u>replete</u> if it is closed under the formation of equivalent objects.) We shall let $c: X^* \longrightarrow X$ be defined to be the coreflection of X into $\underline{Cn}^*$. If X is connected and semilocally 1-connected, then c is the universal covering, but in general, c is a Serre fibration with totally pathwise disconnected fibres. We show that $G(X)$, the fibre of c, has natural group and topological structures defined on it. As a group, $G(X)$ is a quotient of $\pi_1(X)$. One of the topological properties is that $G(X)$ is discrete iff c is a covering (for X locally pathwise connected and T_1). G is functorial both as a group and as a space. Using both of the structures on $G(X)$, we can determine whether or not X is simply connected in the sense of Chevalley (see 4.5, 5.6(b) and 5.7(a))--if X is locally pathwise connected and T_1. For such spaces, we can also obtain some information about the underlying topology of the universal pro-covering of Lubkin, [7]. The universal pro-covering is compared and contrasted with the coreflection X^* in the examples.

We do not know if $G(X)$ is always a topological group, but by 5.2, a large class of topological groups are of the form $G(X)$ for a reasonably well-behaved X. As a corollary to 5.2, we show that if π is any group and N is any normal subgroup of π, there is a connected, locally pathwise connected, T_2 space X with $\pi_1(X) = \pi$, $G(X) = \pi/N$ and $c: X^* \longrightarrow X$ the universal connected covering of X.

Our category theory machinery is based primarily on certain relations between the bicategories of Isbell, [4], and coreflective subcategories. These results can be found in [6] and are stated without proof in the next section. For additional details, see [5] and [6]. The methods used are quite general and could be applied to any one of the large number (i.e. proper class) of coreflective subcategories that lie between the subcategory of simply connected, locally pathwise connected spaces and the subcategory of spaces which are simply connected in the sense of Chevalley. There may also be applications in the category of spaces in the sense of Lubkin [7]. Thus this paper is not exhaustive and is to some degree a preliminary report. In the last section of [6], a closely related coreflection was briefly considered, and analogs of 3.1(a), 3.2(b), 3.6 and its consequences, 3.7, 3.8 and 3.9 were obtained.

Our terminology is based on Freyd, [2]. In particular, a complete category is both left and right complete. We remark that the term "bicategory" has acquired several diverse meanings, but we are using it only as defined below.

Note

For convenience, a coreflective subcategory shall, throughout this paper, be assumed to be full and replete.

2. BICATEGORY STRUCTURES AND COREFLECTIVE SUBCATEGORIES

Definition

Let $\underline{B}$ be a category. Let I and P be classes of morphisms on $\underline{B}$. Then (I,P) is a left bicategory structure on $\underline{B}$ provided that:

LB_0: Every equivalence is in $I \cap P$.

LB_1: I and P are closed under compositions.

LB_2: Every member of I is a monomorphism in $\underline{B}$.

LB_3: Every morphism f can be factored as $f = f_1 f_0$ with $f_1 \in I$ and $f_0 \in P$. If $f = gh$ with $g \in I$ and $h \in P$ is any other factorization, then there exists an equivalence e for which $ef_0 = h$ and $ge = f_1$.

If, in addition to the above, every $f \in P$ is epi, then (I,P) is a bicategory structure.

Definition

Let (I,P) be a left bicategory structure on $\underline{B}$. Then X is an I-subobject of Y if there exists $f\colon X \longrightarrow Y$ with $f \in I$. Similarly, Y is a P-quotient of X if there exists $g\colon X \longrightarrow Y$ with $g \in P$.

2.1 Lemma (Diagonal Property)

Let (I,P) be a left bicategory structure on B. Let $g \varepsilon P$ and $f \varepsilon I$ and morphisms m and n be given with $fm = ng$. Then there exists a morphism r with $rg = m$ and $fr = n$.

2.2 Theorem (Isbell)

Let B be complete and either well-powered or co-well-powered. Let M be the class of all monomorphisms of B and $E^{\#}$ the class of all extremal epimorphisms of B (that is, $e \varepsilon E^{\#}$ iff $e = mg$ and $m \varepsilon M$ imply m is an equivalence). Then $(M, E^{\#})$ is a bicategory structure on B. Moreover, if (I,P) is a left bicategory structure on B, then $E^{\#} \subseteq P$.

Remark

It follows from the diagonal property (2.1), that if $f = f_1 f_0$ with $f_1 \varepsilon M$ and $f_0 \varepsilon E^{\#}$ then f_1 represents the smallest subobject through which f factors.

Definition

Let A be an object of B. Then a morphism $f: X \longrightarrow Y$ has the unique lifting property with respect to A if for every morphism $g: A \longrightarrow Y$ there is a unique $h: A \longrightarrow X$ with $fh = g$.

Definition

Let $\underline{A}$ be a subcategory of $\underline{B}$. We define I(A) to be the set of all morphisms which have the unique lifting property with respect to every object of $\underline{A}$.

We define $P(\underline{A})$ to be the class of all morphisms p for which $p = ig$ and $i \in I(A)$ imply i is an equivalence.

2.3 and 2.4, stated below, are the main results of [6].

2.3 Theorem

Let $\underline{C}$ be a complete, well-powered and co-well-powered category. Let $\underline{A}$ be any full subcategory. Let $\underline{B}$ be the full subcategory of all quotients of coproducts of members of $\underline{A}$. Working in the category $\underline{B}$, let $I = I(\underline{A})$ and $P = P(\underline{A})$ be defined as above. Let $\underline{A}^*$ be the subcategory of all P-quotients of coproducts of members of $\underline{A}$. Then:

(a) $\underline{A}$ is coreflective in $\underline{C}$ iff $\underline{A} = \underline{A}^*$ and $\underline{B}$ is "I-wellpowered" (i.e., if each object of $\underline{B}$ has a representative set of I-subobjects). If $\underline{B}$ is I-wellpowered then:

(b) (I,P) is a left bicategory structure on $\underline{B}$.

(c) $\underline{A}^*$ is the smallest coreflective subcategory of $\underline{C}$ for which $\underline{A} \subset \underline{A}^*$.

2.4 Proposition

Let $\underline{A}$, $\underline{A}^*$, $\underline{B}$, $\underline{C}$, and (I,P) be as above. Assume that

$\underline{B}$ is I-well-powered. Then the following statements are equivalent:

(a) $A \in \underline{A}^*$

(b) $A \in \underline{B}$ and whenever $f: X \longrightarrow A$ is in I then f is an equivalence.

(c) $A \in \underline{B}$ and every $\underline{B}$ morphism $f: X \longrightarrow A$ is in P.

(d) $A \in \underline{B}$ and every member of I has the unique lifting property with respect to A.

3. CONSTRUCTION OF THE COREFLECTION

We now apply the results of the previous section to the study of the smallest coreflective subcategory containing all contractible spaces (i.e. spaces which are homotopically equivalent to a point). Working in the category of pointed topological spaces, we shall let $\underline{Cn}$ be the subcategory of all contractible spaces. The subcategory $\underline{PC}$, of all pathwise connected spaces is clearly the category of all "quotients" (i.e. continuous images) of coproducts of members of $\underline{Cn}$.

Let $I = I(\underline{Cn})$ and $P = P(\underline{Cn})$ be defined as in the previous section. It is shown in [5, p. 409] that $\underline{PC}$ is well-powered, hence I-well-powered and so 2.3 and 2.4 apply. That is, (I,P) is a left bicategory system on $\underline{PC}$ which can be used to compute $\underline{Cn}^*$, the coreflective subcategory generated by $\underline{Cn}$.

For each $X \in \underline{PC}$, we shall construct the coreflection map $c : X^* \longrightarrow X$ associated with $\underline{Cn}^*$. As explained in the introduction, this map is a generalized universal simply connected covering of X. We shall also discuss some elementary properties of c, $\underline{Cn}^*$ and I.

Note

(1) From here on every space considered shall be assumed to be in PC.

(2) From here on the term "quotient" shall be used in its topological rather than its categorical sense.

Notation

In this paper the term simply connected shall be used as in [3], that is to denote a member of PC for which all loops shrink to a point. The term covering projection (or just covering) shall be used as in [9], without any restrictions on the base space, such as local connectedness. A Chevalley simply connected space shall be a connected, locally connected space with no nontrivial connected coverings. (We drop Chevalley's blanket assumption, in [1], that a space be T_2.) The terms Serre Fibration and Hurewicz Fibration shall have their usual meanings. Note that the unmodified terms "fibering", "fibration", etc., have the former sense in [3] and the latter sense in [9].

3.1 Proposition

(a) Every member of I is a Serre fibration with totally pathwise disconnected fibres.

(b) Every Hurewicz fibration, in PC, with totally pathwise disconnected fibres, is in I.

Proof. We shall postpone the proof of (a) until the end of this section. As for (b), let $f: (Y,y_0) \longrightarrow (X,x_0)$ be a Hurewicz fibration with totally pathwise disconnected fibres. Let $(A,a_0) \in Cn$ and $g: (A,a_0) \longrightarrow (X,x_0)$ be given. Let $h_t: A \longrightarrow A$ be a homotopy with $h_0 = a_0$ (the constant map) and h_1 the identity map on A. By the covering homotopy property, there exists a homotopy $n_t: A \longrightarrow Y$ such that $n_0 = y_0$ (the constant map) and $fn_t = gh_t$. Then the path $(n_t(a_0))$ lies over $g(h_t(a_0))$, a shrinkable loop. It readily follows that $n_1(a_0) = n_0(a_0) = y_0$ and so $n_1: (A,a_0) \longrightarrow (Y,y_0)$ is the desired lifting of g.

3.2 Corollary

(a) Every simply connected space, which is the quotient of a contractible space, is in Cn*. Thus, every simply connected and locally pathwise connected space is in Cn*.

(b) Every locally pathwise connected member of Cn* is Chevalley simply connected.

<u>Proof</u>. (a) Let X be simply connected and let $q: A \longrightarrow X$ be a quotient map with $A \in \underline{Cn}$. Let $C: X^* \longrightarrow X$ be the coreflection map. Then there exists h with $ch = q$ which implies that c is a quotient map. But it is easily shown that c is one-one, hence, c is a homeomorphism. Finally every connected, locally pathwise connected space is the quotient of a contractible space, namely the path space defined below, see [9, p. 75].

(b) As shown in [9, p. 67], every covering is a Hurewicz fibration. Thus, if X is locally pathwise connected and in $\underline{Cn}^*$, then every connected covering of X is pathwise connected, hence in I, hence trivial by 2.4(b).

<u>Definition</u>

Let $(X, X_0) \in \underline{PC}$ be given. Then $P(X, x_0)$, the <u>path space over X</u> is the set of maps from $([0, 1], 0) \longrightarrow (X, X_0)$ together with the compact-open topology. The base point of $P(X, x_0)$ is the constant map.

We let $\phi : P(X, x_0) \longrightarrow X$ be the map for which $\phi(f) = f(1)$.

When there is no danger of confusion, we shall use $P(X)$ instead of $P(X, x_0)$.

3.3 Theorem

Let ϕ: P(X) $\longrightarrow$ X be as above. Factor ϕ as $\phi = c\phi_0$ where c is mono in PC and ϕ_0 is an extremal epimorphism in PC (we are applying 2.2 to PC). Let ϕ_0: P(X) $\longrightarrow X^*$ and c: $X^* \longrightarrow$ X.

Then $X^* \in Cn^*$ and c is the coreflection map associated with X.

Proof: We first note that $\phi_0 \in P$ as $E^\# \subseteq P$. Then $X^* \in Cn^*$ as P(X) $\in$ Cn (see [9, p. 75]) and Cn^* is closed under P-quotients. It remains to show that c $\in$ I. The argument given in 3.1 can be used to show that every member of Cn "lifts" via c. Moreover, the lifting is unique as c is mono in PC.

3.4 Corollary

Let ϕ: P(X) $\longrightarrow$ X and c: $X^* \longrightarrow$ X be as above. If X is T_1, then $c^{-1}(x)$ is a totally pathwise disconnected quotient of $\phi^{-1}(x)$ for all x $\in$ X.

3.5 Corollary

Let X be T_1 and let c: $(X^*, x_0^*) \longrightarrow (X, x_0)$ be the coreflection. Let h: [0, 1] $\longrightarrow$ X be a path with $h(0) = x_0$ and $h(1) = x_1$. Then the map from $c^{-1}(x_0)$ to $c^{-1}(x_1)$ induced by h (which sends y_0 to y_1 iff there exists

$\bar{h}$: $[0, 1] \longrightarrow X^*$ with $c\bar{h} = h$, $\bar{h}(0) = y_0$ and $\bar{h}(1) = y_1$) is a homeomorphism. This homeomorphism depends only on the homotopy class of h.

Proof: Consider the well known continuous map from $\phi^{-1}(x_0)$ to $\phi^{-1}(x_1)$ which sends the loop λ to the path $h + \lambda$ (using the usual sum for paths having an endpoint in common). Taking quotients, we obtain the desired map from $c^{-1}(x_0)$ to $c^{-1}(x_1)$. This map has an obvious inverse and also clearly remains invariant under homotopy.

3.6 Proposition (Base Point Change)

Let $f: (Y, y_0) \longrightarrow (X, x_0)$ be in I. Let $y_1 \in Y$ be arbitrary and let $x_1 = f(y_1)$. Then $f: (Y, y_1) \longrightarrow (X, x_1)$ is also in I.

Proof: Let $\bar{h}: [0, 1] \longrightarrow Y$ be such that $\bar{h}(0) = y_0$ and $\bar{h}(1) = y_1$. Let $h = f\bar{h}$. Let $(A, a_1) \in \underline{Cn}$ and $g: (A, a_1) \longrightarrow (X, x_1)$ be given. Let $Z = ([0, 1], 1) + (A, a_1)$ be the coproduct in $\underline{PC}$ (i.e. the wedge product). Then $h + g: Z \longrightarrow (X, x_1)$ is well defined if h is regarded as mapping $([0, 1], 1)$ into (X, x_1). We can also regard $h + g$ as mapping $(Z, 0)$ into (X, x_0). Clearly $Z \in \underline{Cn}$ and since $f \in I$ there exists $\phi: (Z, 0) \longrightarrow (Y, y_0)$ with $f\phi = h + g$. It follows that $\phi|[0, 1] = \bar{h}$, hence $\phi(1) = y_1$ and thus, $\phi|A$ is the desired lifting of g.

3.7 Corollary

Let $X \in \underline{PC}$ and $x_0, x_1 \in X$ be given. Then $(X, x_0) \in \underline{Cn}^*$ iff $(X, x_1) \in \underline{Cn}^*$.

Proof: Follows in view of 2.4(b) and the above.

3.8 Corollary

Let $c: (X^*, x_0^*) \longrightarrow (X, x_0)$ be the coreflection of (X, x_0). Let $x_1^* \in X^*$ be arbitrary and let $x_1 = c(x_1^*)$. Then $c: (X^*, x_1^*) \longrightarrow (X, x_1)$ is the coreflection of (X, x_1).

Remark

In view of the above results, we shall freely continue our practice of frequently suppressing any mention of the base point.

Definition

Let $c: X^* \longrightarrow X$ be the coreflection of X. Then the deck translation group for X is the group of all automorphisms $g: X^* \longrightarrow X^*$ such that $cg = c$. (These automorphisms are not required to preserve any base points.)

3.9 Theorem

Let $c: X^* \longrightarrow X$ be the coreflection of X. Then the deck translation group acts transitively on each fibre of c so that whenever $x_1, x_2 \in X^*$ are given with $c(x_1) = c(x_2)$, there is a unique deck translation g for which $g(x_1) = x_2$.

Proof: Assume that $c(x_1) = c(x_2) = x_0$. Then $c: (X^*, x_1) \longrightarrow (X, x_0)$ and $c: (X^*, x_2) \longrightarrow (X, x_0)$ are both coreflections of (X, x_0) in view of 3.7. Thus there are maps $g: (X^*, x_1) \longrightarrow (X, x_2)$ and $h: (X^*, x_2) \longrightarrow (X^*, x_1)$ with $cg = c$ and $ch = c$. Clearly h is the inverse of g and so g is a deck translation. The uniqueness of g follows since c is mono in PC.

Proof of 3.1 (a): Let $f: Y \longrightarrow X$ be in I. Let $S \varepsilon$ Cn be arbitrary and let $g: S \longrightarrow S \times [0, 1]$ be any map. Let $m: S \longrightarrow Y$ and $n: S \times [0, 1] \longrightarrow X$ be such that $fm = ng$. (In view of 3.6, we can regard this as a diagram of pointed spaces by choosing any base point in S, etc.) Note that $g \varepsilon P$ by 2.4(c) and so by 2.1, there exists a map r with $rg = m$ and $fr = n$. This clearly implies that f has the covering homotopy property with respect to all contractible spaces and so f is a Serre fibration.

4. TOPOLOGICAL AND GROUP STRUCTURES ON THE FIBRE

Definition

Let $c: (X^*, x_0^*) \longrightarrow (X, x_0)$ be the coreflection of (X, x_0). We define $G(X, x_0) = c^{-1}(x_0)$ and let $n: \pi_1(X, x_0) \longrightarrow G(X, x_0)$ be the natural function. [Notice that $\pi_1(X, x_0)$ and $G(X, x_0)$ are both equivalence classes of loops in $\phi^{-1}(x_0)$, see 3.3. Since $G(X, x_0)$ is totally pathwise disconnected, homotopic loops must be equivalent in $G(X, x_0)$. Hence we can readily obtain the above function n.]

4.1 Proposition

$G(X, x_0)$ can be given a group structure so that $n: \pi_1(X, x_0) \longrightarrow G(X, x_0)$ is a group homomorphism. Then $G(X, x_0)$ is canonically isomorphic to the deck translation group.

Proof: For each $x \in G(X, x_0)$ associate the deck translation g for which $g(x_0^*) = x$. By 3.9, this places $G(X, x_0)$ in one-one correspondence with the deck translation group. If we give $G(X, x_0)$ the induced group structure, it is then easy to show that n is a group homomorphism.

Definition

Let $n: \pi_1(X, x_0) \longrightarrow G(X, x_0)$ be the above homomorphism. We define $N(X, x_0)$ to be the kernel of n.

4.2 Proposition:

π_1, G and N are functors from PC into groups. The action of these functors is, to within an equivalence, independent of the base point. We can regard n as a natural transformation from π_1 to G. Finally N(X) is isomorphic to $\pi_1(X^*)$.

Proof: Straightforward.

Remarks

$G(X, x_0)$ is not only a group, but is also a topological space (as a subspace of X^*). Perhaps it is a topological group, although so far we can only show that the "algebraic" functions of one variable (such as the function sending x into $xax^{-1}b$ for a, b fixed) are continuous for X a T_1 space. This fact follows since $G(X, x_0)$ is a quotient of the loop space $\phi^{-1}(x_0)$.

It can easily be shown that G is a functor from PC to the category of groups with topologies.

If X is semilocally 1-connected (see [9, p. 78]) then $c: X^* \longrightarrow X$ is the universal simply connected covering, $n: \pi_1(X) \longrightarrow G(X)$ is a natural isomorphism and G(X) has the discrete topology.

Definition

X is semilocally G-connected if each $x \in X$ has a neighborhood U such that the natural map $G(U) \longrightarrow G(X)$ is trivial.

4.3 Theorem

Let X be locally pathwise connected and T_1 and let $c: X^* \to X$ be its coreflection. Then the following statements are equivalent:

(a) $c^{-1}(x)$ is discrete for some (and hence for all) $x \in X$.

(b) X is semilocally G-connected.

(c) c is a covering of X (in which case c is the universal connected covering of X).

Proof: (a) $\Longrightarrow$ (b): Let $x_0 \in X$ be given and choose $x_0^* \in X^*$ such that $c(x_0^*) = x_0$. Let W be an open subset of X^* such that $W \cap p^{-1}(x_0) = \{x_0^*\}$. Let ϕ_0: $P(X, x_0) \longrightarrow (X^*, x_0)$ be as in section 3, then $\phi_0^{-1}(W)$ is open in the compact-open topology. Thus, there are non-empty compact sets $K_1,\ldots,K_n$ contained in $[0, 1]$ and open sets $U_1,\ldots,U_n$ of X, such that $k \in \cap M(K_i, U_i) \subseteq \phi_0^{-1}(W)$, where k is the constant path on x_0 and $M(K, U) = \{f \in P(X) | f(K) \subseteq U\}$. Let $U \subseteq \cap U_i$ be a pathwise connected neighborhood of x_0. Then any loop λ, starting at x_0 and with range in U, is in $\cap M(K_i, U_i)$. Thus $\phi_0(\lambda) \in W \cap c^{-1}(x_0)$ and so $\phi_0(\lambda) = x_0^*$. It is now easy to show that $G(U) \longrightarrow G(X)$ is trivial.

(b) $\Longrightarrow$ (c): Let U be an open, pathwise connected neighborhood of a given $x \in X$ such that $G(U) \longrightarrow G(X)$ is trivial. Let $\{V_\alpha\}$ be the path components of $c^{-1}(U)$. It is easily shown that each restriction $c|V_\alpha$ is one-one. In view of the deck translations, the V_α's are canonically homemorphic. It can further be shown that each $c|V_\alpha$ is a quotient map using the fact that $c|UV_\alpha \longrightarrow U$ is a quotient map. (c is a quotient map as ϕ is, see [9, p. 75], and $c\phi_0 = \phi$. Thus $c|c^{-1}(U)$ is also a quotient map as U is open.) It follows that each $c|V_\alpha$ is a homeomorphism. Thus c is a covering. It is clearly a universal connected covering in view of 2.4(d) as every other such covering is in I.

(c) $\Longrightarrow$ (a): Trivial.

<u>Definition</u>: Let X be T_1. Let $f: (Y, y_0) \longrightarrow (X, x_0)$ be in I. Then f is <u>regular</u> if for each $y \in f^{-1}(x_0)$ there exists an automorphism $g: (Y, y_0) \longrightarrow (Y, y)$ such that $fg = f$. Equivalently, $f: (Y, y_0) \longrightarrow (X, x_0)$ in I is regular iff the map $g: (X^*, x_0^*) \longrightarrow (Y, y_0)$ for which $fg = c$ (which exists as $f \in I$ and by 2.4(d)) is such that $g^{-1}(y_0)$ is a normal subgroup of $G(X, x_0)$. (Thus it follows that our definition agrees with that on p. 74 of [9].)

Clearly, regularity is independent of the base point.

<u>4.4 Corollary</u>

Let X be locally pathwise connected and T_1. Then there is a canonical one-one correspondence between regular coverings of X and open normal subgroups of G(X).

<u>Proof</u>: Let H be an open normal subgroup of G(X). Choose any base point in X^*. Then G(X) can be regarded as the deck translation group. Consider the quotient space X^*/H ob-

tained from X^* by identifying x with h(x) whenever the deck translation h is (regarded as) a member of H. (The space X^*/H does not depend on our choice as H is normal.)

The map $f: X^*/H \longrightarrow X$ induced by c can be shown to be a covering by means of the arguments given in 4.3.

4.5 Corollary

Let X be locally pathwise connected and T_1. Then X is Chevalley simply connected iff G(X) has no proper open normal subgroups.

Proof: For (connected and) locally pathwise connected X, the coverings of X are in effective one-one correspondence with the coverings in the sense of Lubkin, [7], (see the remark below). From Theorem 1 on p. 211 of [7] it follows that any non-trivial connected covering of X gives rise to a non-trivial regular connected covering.

Remark

In what follows, we shall refer to the generalized uniform spaces of Lubkin (called spaces in [7]) as Lubkin spaces. The main facts that we shall use, from [7], are that each topological space is, in a natural way, a Lubkin space and every Lubkin space has an underlying topology. Moreover, if X is (connected and) locally connected, then each covering $f: Y \longrightarrow X$ is such that Y can be regarded as a compatible

Lubkin space and f as a Lubkin covering. Conversely every Lubkin covering of such an X is of this type.

The universal pro-covering, $\tilde{X}$, of X, is the inverse limit of all connected Lubkin coverings of X. We shall let $\overline{X}$ denote the underlying topology of $\tilde{X}$. If X is (connected and) locally pathwise connected, then $\overline{X}$ is the inverse limit of all regular coverings of X (in view of Theorem 1 on p. 211 and Corollary 1.4 on p. 215 of [7].) It follows that X is Chevalley simply connected iff $X = \overline{X}$.

The relationship between X^* and $\overline{X}$ (for connected, locally pathwise connected, X) is given by:

4.6 Theorem

Let X be (connected and) locally pathwise connected and T_1. Let $f: \overline{X} \longrightarrow X$ be the underlying topology and continuous function associated with the universal pro-covering of X.

Then there exists $\overline{c}: X^* \longrightarrow \overline{X}$ such that $f\overline{c} = c$. The map $\overline{c}$ is the coreflection of $\overline{X}$ and $G(\overline{X})$ is the intersection of all open normal subgroups of G(X). Moreover $\overline{X}$ is the quotient space $X^*/G(\overline{X})$.

Proof: The existence of $\overline{c}$ follows from 2.4(d) as f is clearly in I. A direct proof shows that $\overline{c}$ is the coreflection and that $G(\overline{X})$ is the above intersection. It can readily be shown that $\bar{X}$ is locally pathwise connected and so $\bar{c}$ is a quotient map (see 3.2).

Remark

Examples 5.6 (b), 5.7, and 5.8 pertain to the relationship between $\bar{X}$ and X^*.

5. EXAMPLES

5.1 Lemma

Let X be T_1. Let $\phi\colon P(X) \longrightarrow X$ be factored as $\phi = mq$ where $q\colon P(X) \longrightarrow Q$ is a quotient map and $m\colon Q \longrightarrow X$ is mono in PC. Suppose that for each $x \in X$, $m^{-1}(x)$ is the largest totally pathwise disconnected quotient of $\phi^{-1}(x)$. Then $Q = X^*$ and $m = c$.

Proof: By 3.3, we have that X^* is the smallest "subobject" of X through which ϕ factors. Thus there exists $f\colon X^* \longrightarrow Q$ such that $mf = c$. In view of the quotient hypothesis and 3.4, we obtain a map $h\colon Q \longrightarrow X^*$ with $hq = \phi_0$. It is easy to show that h is the inverse of f and hence, $Q = X^*$ and $m = c$.

5.2 Theorem:

Let G be a T_0 topological group. Let $\{H_\alpha\}$ be a base for the neighborhoods of the identity element such that each H_α

is an open subgroup. Then there exists a (connected and) locally pathwise connected T_2 space X such that G(X) is topologically and algebraically equivalent to G.

<u>Proof</u>: Let (S, s_0) be a pathwise connected, locally simply connected pointed T_2 space such that $\pi_1(S, s_0) = G$. For each α let $S_\alpha \subseteq S$ be such that $S_\alpha \longrightarrow S$ induces an injection $\pi_1(S_\alpha) \longrightarrow \pi_1(S)$ such that $\pi_1(S_\alpha)$ maps onto H_α. (Such an S can be constructed by taking a wedge of 1-spheres and attaching some 2-cells, as in [9, p. 143-148].)

[A straightforward attempt to construct X would involve a change of the topology on S--requiring that a neighborhood of s_0 must contain some S_α. This approach would work except for the fact that S, when retopologized, might acquire some new loops.]

Let R be the real line. We define the underlying set of X as $S \times R$ with $\{s_0\} \times R$ identified to a single point called x_0. Let $e: S \times R \longrightarrow X$ be the canonical map. We topologize X so that $U \subseteq X$ is open iff $e^{-1}(U)$ is open in the product topology and, for $x_0 \in U$, we also require that there exist α and a countable subset $C \subseteq R$ such that $S_\alpha \times (R \setminus C) \subseteq e^{-1}(U)$.

We let $r: X \longrightarrow S$ be the obvious projection function. r is not continuous at x_0 but r does preserve the convergence of sequences. Thus, if $f: [0, 1] \longrightarrow X$ is continuous , so is rf: $[0, 1] \longrightarrow S$. Similarly, if f, g are homotopic paths on X, then so are rf and rg. Hence $\pi_1(X) = \pi_1(S) = G$.

Let $\phi^{-1}(x_0)$ be the loop space of X (as in section 3) and let $v\colon \phi^{-1}(x_0) \longrightarrow \pi_1(X, x_0)$ take each loop into its equivalence class. We claim that the quotient topology on $\pi_1(X, x_0)$ induced by v, is homeomorphic to the given topology on G. It suffices to look at neighborhoods of the identity (see remark following 4.2).

Let U be a neighborhood of the identity in the quotient topology. Then there exist compact sets $K_1, \ldots, K_n \subseteq [0, 1]$ and open subsets $U_1, \ldots, U_n$ of X such that $k \in \cap M(K_i, U_i) \subseteq v^{-1}(U)$, where k is the constant loop at x_0. Clearly there exists an S_α and an $t \in R$ with $e(S_\alpha \times \{t\}) \subseteq \cap U_i$, which implies $H_\alpha \subseteq U$. Thus U is a neighborhood in the given topology.

It remains to show that each H_α is open in the quotient topology. Let H_α and $h \in v^{-1}(H_\alpha)$ be given. We must find compact sets $K_1, \ldots, K_n$ and open sets $U_1, \ldots, U_n$ such that $h \in \cap M(K_i, U_i) \subseteq v^{-1}(H_\alpha)$.

Let the open set $[0, 1] \setminus h^{-1}(x_0)$ be written as the disjoint union, $\cup I_i$, of open intervals. Choose any simply connected neighborhood W of s_0 (in the space S). Then $rh(I_i) \subseteq W$ for all but finitely many I_i. [If not, let $\{x_n\}$ be a sequence such that $rh(x_n) \notin W$, for all n, and each I_i contains at most one x_n. Then h separates $\{x_n\}$ from its cluster points.]

Let $I_1, I_2, \ldots, I_m$ be the intervals (arranged in the obvious increasing order) for which $rh(I_i)$ is not a subset of W. Let a_i be the loop effectively defined as $h|I_i$ for $i = 1, \ldots, m$. Let $\bar{a}_i = v(a_i)$. Choose H_β so that $H_\beta \bar{a}_1 H_\beta \bar{a}_2 H_\beta \ldots H_\beta \bar{a}_n H_\beta \subseteq H_\alpha$. [This is possible since the map sending x into $x\bar{a}_1 x\bar{a}_2 x \ldots x\bar{a}_n x$ is continuous, sends the identity into $\bar{a}_1 \bar{a}_2 \ldots \bar{a}_n = v(h) \in H_\alpha$, hence sends some neighborhood of the identity into a subset of H_α.] Let K_1 be a compact neighborhood of the complement of $I_1 \cup I_2 \cup \ldots \cup I_m$ such that $rh(K_1) \subseteq W$. Let U_1 be an open neighborhood of x_0 such that $r(U_1) \subseteq W$ and $\pi_1(U_1) = H_\beta$. Choose compact intervals $K_2, \ldots, K_n$ and open sets $U_2, \ldots, U_n$ so that $I_1 \cup I_2 \cup \ldots \cup I_m \subseteq K_1 \cup \ldots K_n$, that each U_i is simply connected, for $2 \le i \le n$, and that $h(K_i) \subseteq U_i$. Then $\cap M(K_i, U_i)$ is the desired neighborhood of h.

We let $q\colon P(X) \longrightarrow Q$ be the quotient map which identifies homotopic paths. Let $m\colon Q \longrightarrow X$ be such that $mq = \phi$. In view of 5.1, it suffices to show that m is mono in <u>PC</u>. This easily follows from the claim that f = g whenever mf = mg and $f, g\colon [0, 1] \longrightarrow Q$. To prove this claim, we need only show that f and g agree on an open and closed subset of [0, 1] (in addition to agreeing at the base point).

We shall show for each $t \in [0, 1]$ that either f = g or $f \neq g$ in a neighborhood of t. Let $x = mf(t) = mg(t)$. We first assume that $x \neq x_0$. Let $U \subseteq X$ be any simply connected neighborhood of x. It is readily shown that the underlying set of $m^{-1}(U)$ can be regarded in a natural way as $U \times G$. Moreover, the subsets $U \times H_\alpha$ can be shown to be open in $m^{-1}(U)$ by means of the above argument. Thus, the path components of $m^{-1}(U)$ are of the form $U \times \{g\}$ and hence either $f = g$ or $f \neq g$ in the set $(mf)^{-1}(U)$. If $x = x_0$, the same argument works if U is a simply connected sequential neighborhood of x_0 (that is, no sequence outside of U can converge to x_0).

5.3 Corollary

Let π be a topological group (not necessarily T_0). Let $\{H_\alpha\}$ be a base for the neighborhoods of the identity such that each H_α is an open subgroup. Let $N = \cap H_\alpha$. (It then follows that N is normal.) Let $G = \pi/N$ have the quotient topology. Then there exists a connected, locally pathwise connected T_2 space X with $\pi_1(X) = \pi$, $N(X) = N$, and $G(X) = G$.

Proof: Construct X as in the above proof using π in place of G. The same arguments then apply.

5.4 Corollary

If π is any group and N is any normal subgroup, then there exists a semilocally G-connected T_2 space X such that $\pi_1(X) = \pi$, $N(X) = N$ and $G(X) = \pi/N$, with the discrete topology.

Thus there exist universal connected covering spaces $c: X^* \longrightarrow X$ with $\pi_1(X)$ any preassigned group and $\pi_1(X^*)$ equivalent to any preassigned normal subgroup.

5.5 Corollary

Let π be any group and let $\{H_\alpha\}$ be any collection of subgroups such that:

(1) For all α, β, there is a λ with $H_\lambda \subseteq H_\alpha \cap H_\beta$.

(2) For each α and each $g \in \pi$ there exists β such that $g^{-1}H_\beta g \subseteq H_\alpha$.

Then there exists a connected, locally pathwise connected T_2 space X such that $\pi_1(X) = \pi$, $N(X) = N = \cap H_\alpha$ and $G(X) = \pi/N$ with $\{H_\alpha/N\}$ a base the neighborhoods of the identity.

5.6 Example

(a) There exists $X \in \underline{Cn}^*$ such that X is not simply connected (in fact $\pi_1(X)$ can be preassigned).

(b) There exists a Chevalley simply connected and locally pathwise connected T_2 space X for which G(X) is non-trivial (hence $X \notin \underline{Cn}^*$).

Proof: (a) Follows from 5.4 with $\pi = N$.

(b) Let Z^+ be the positive integers. Let G be the group of all permutations of Z^+. For each $n \in Z^+$, let H_n be the subgroup of all $f \in G$ such that $f(i) = i$ for all $i \leq n$. We claim that (1) and (2) of 5.5 are satisfied by $\{H_n\}$.

Clearly (1) holds. As for (2) let $g \in G$ and $n > 0$ be given. Choose m so that $m \geq g(i)$ for all $i \leq n$. Then $g^{-1}H_m g \subseteq H_n$. Thus there is a suitable X with $\pi_1(X) = G(X) = G$ and such that $\{H_n\}$ is a base of open neighborhoods of the identity.

We claim that G has no proper normal open subgroups which by 4.5, implies (b). Assume that H is a normal subgroup and that $H_n \subseteq H$ for some n. Note that if $f \in G$ fixes any set of n elements, then $f \in g^{-1}H_n g$ for some g and so $f \in H$. Now let $s \in G$ be arbitrary. (It suffices to show $s \in H$.) Let $\{0_\alpha\}$ be the collection of orbits of s (an orbit being a minimal set 0_α such that $s(0_\alpha) = 0_\alpha$). If s has infinitely many orbits, then $s = fg$ where f restricts to the identity on "half" of the orbits and g restricts to the identity on the "other half". f and g thus fix infinitely many elements and so $f, g \in H$ implying $s \in H$. Finally assume that s has only a finite number of orbits. Then s must have at least one infinite orbit U. Let $u \in U$ be given. It follows that $s^i(u) \neq s^j(u)$ for $i \neq j$. Define $g\colon Z^+ \longrightarrow Z^+$ so that $g(v) = v$ for $v = s^1(u), s^2(u), \ldots, s^n(u)$. Further define $g(u) = s^{n+1}(u)$ and $g(z) = s(z)$ for all other values. Let $f = sg^{-1}$. Clearly g fixes at least n elements and f fixes infinitely many elements so $f, g \in H$. Thus $s = fg \in H$.

5.7 Example

(a) There exists a locally pathwise connected T_2 space

X such that $\bar{X}$ is not Chevalley simply connected. (Equivalently, $\bar{\bar{X}} \neq \bar{X}$.)

(b) The above space X has a universal connected covering which is not in Cn.*

Proof: For this example, let G be the group of all permutations $g: Z \longrightarrow Z$ such that $g(-n) = -g(n)$ for all integers n. Let A be the subgroup of all $a: Z \longrightarrow Z$ with $|a(n)| = |n|$ for all n. Let H_n be the subgroup of all $h: Z \longrightarrow Z$ such that $h(i) = i$ for $i = 1, 2, \ldots, n$. Topologize G so that $\{A \cap H_n\}$ is a base for the neighborhoods of the identity. It is readily verified that there exists a suitable X with $G(X) = G$.

It is also easily shown (using arguments similar to those in 5.6) that A is the smallest open normal subgroup of G. Hence, by 4.6, we have $G(\bar{X}) = A$. But A is abelian and has many open normal subgroups. Thus $\bar{X}$ is not Chevalley simply connected (in fact $\bar{\bar{X}} = X^*$). Finally the map $f: \bar{X} \longrightarrow X$ is a covering in view of 4.4. By construction of $\bar{X}$ this is the universal connected covering.

Remarks

(1) In the above example, it is easily shown that the Lubkin universal pro-covering $\tilde{X} \longrightarrow X$ is actually the universal Lubkin covering of X. Since Lubkin coverings compose, the Lubkin space $\tilde{X}$ has no non-trivial connected coverings. On the

other hand, we have seen that $\overline{X}$ has many non-trivial connected coverings. Thus $\overline{X} \neq \tilde{X}$ which means $\tilde{X}$ cannot be a topological space.

(2) It should be mentioned that the universal pro-covering is not always a quotient of X^* if X is not locally pathwise connected. There exist spaces X with $\pi_1(X) = 0$, but $\tilde{X} \longrightarrow X$ not one-one (see examples 4, 7, and 22 of [7]).

5.8 Example

Let $X = R \times R \setminus \{(2^{-n}, 0)\}$. This space is discussed in [7, p. 232] and in [6]. It can be shown that the universal procovering, $\overline{X}$, is simply connected and hence, $\overline{X} = X^*$. It follows from [7, p. 232], that the group G(X) is the completion of the free group with countably many generators and open subgroups containing all, but finitely many generators. As a group G(X) $= \pi_1(X)$, but there may be occasions when the above description of G(X) is easier to work with than the algebraic description of $\pi_1(X)$ as a free group with uncountably many generators.

REFERENCES

[1] Chevalley, C., *Theory of Lie groups, I*, Princeton University Press, Princeton, (1946).

[2] Freyd, P., *Abelian categories*, Harper and Row, New York, (1964).

[3] Hu, S. T., *Homotopy theory*, Academic Press, New York, (1959).

[4] Isbell, J. R., "Some remarks concerning categories and subspaces", *Canad. J. Math.*, 9; pp. 563-577, (1957).

[5] Kennison, J. F., "A note on reflection maps", *Illinois J. Math.*, 11; pp. 404-409, (1967).

[6] ________, "Full reflective subcategories and generalized covering spaces", *Illinois J. Math.*, (to appear).

[7] Lubkin, S., "Theory of covering spaces", *Trans. Amer. Math. Soc.*, 104; pp. 205-238, (1962).

[8] MacLane, S., "Categorical algebra", *Bull. Amer. Math. Soc.* 71; pp. 40-106, (1965).

[9] Spanier, E. H., *Algebraic topology*, McGraw-Hill, New York, (1966).

DEDUCTIVE SYSTEMS AND CATEGORIES

II. STANDARD CONSTRUCTIONS AND CLOSED CATEGORIES*

by

Joachim Lambek**

0. INTRODUCTION

We wish to explore the connection between

(1) pre-ordered sets with structure,

(2) deductive systems,

(3) categories with structure.

By a pre-ordered set we mean a set with a transitive and reflexive, but not necessarily anti-symmetric, relation $\leq$. It would be premature to give a rigorous definition of what we have in mind by "structure" and "deductive system". Instead we shall illustrate with an example.

(1) A semilattice (with largest element) is a triple $(\underline{A},T,\wedge)$, where $\underline{A}$ is a pre-ordered set, T an

---

* Familiarity with Deductive Systems and Categories (I) is not presumed. Thanks are due to the Battelle Institute for inviting me to give these lectures and to S. Eilenberg for his stimulating comments.

** Mathematics Dept., McGill University, Montreal

element of $\underline{A}$, and $\wedge$ a binary operation on $\underline{A}$ such that

$$A \leq T,$$

$$C \leq A \wedge B \Longleftrightarrow C \leq A \ \& \ C \leq B.$$

(2) For any discrete set $\underline{X}$ there is a deductive system $D(\underline{X})$ described as follows:

Its terms are all elements X of $\underline{X}$, also T is a term; finally if A and B are terms, so is $A \wedge B$. (It is understood that nothing else is a term.)

Its formulas are all expressions of the form $A \longrightarrow B$ where A and B are terms.

Its axioms (or axiom schemes) and rules of inference are listed as follows:

$$1: \ A \longrightarrow A, \quad 2: \ \frac{A \longrightarrow B \quad B \longrightarrow C}{A \longrightarrow C}, \quad 3: \ A \longrightarrow T,$$

$$4: \ \frac{C \longrightarrow A \wedge B}{C \longrightarrow A}, \quad 5: \ \frac{C \longrightarrow A \wedge B}{C \longrightarrow B}, \quad 6: \ \frac{C \longrightarrow A \quad C \longrightarrow B}{C \longrightarrow A \wedge B}.$$

Proofs are written in tree-form. For example, the following is a proof:

$$6\ \frac{5\ \dfrac{A \wedge B \xrightarrow{1} A \wedge B}{A \wedge B \longrightarrow B} \qquad 4\ \dfrac{A \wedge B \xrightarrow{1} A \wedge B}{A \wedge B \longrightarrow A}}{A \wedge B \longrightarrow B \wedge A}$$

Any formula appearing as a proof, in particular the last

formula, is called provable or a theorem. Thus we see from the above that $A \wedge B \longrightarrow A \wedge B$, $A \wedge B \longrightarrow B$, $A \wedge B \longrightarrow A$, and $A \wedge B \longrightarrow B \wedge A$ are all theorems, but only the first is an axiom.

(3) A cartesian category is a quintuple $(\underline{A}, T, \wedge, \alpha, \beta)$, where $\underline{A}$ is a category, T is an object of $\underline{A}$, $\wedge: \underline{A} \times \underline{A} \longrightarrow \underline{A}$ is a bifunctor, and α and β are natural isomorphisms as follows:

$$\alpha(A): \underline{1} \longrightarrow [A,T] ,$$
$$\beta(A,B,C): [C,A] \times [C,B] \longrightarrow [C,A \wedge B] .$$

Here $\underline{1}$ is "the" one-element set, and $[A,B]$ is short for $\mathrm{Hom}_{\underline{A}}(A,B)$. Thus T is "the" terminal object of $\underline{A}$ and $\wedge$ is the usual categorical product.

We shall see later that the deductive system $D(\underline{X})$ can be used to construct not only the free semilattice, but also the free cartesian category generated by $\underline{X}$. In the first case we are only interested in the theorems of the deductive system, but in the second case the proofs will play an important role.

More generally one can form a deductive system $D(\underline{X})$ when $\underline{X}$ is a pre-ordered set by adjoining the axiom

$$0: X \longrightarrow Y$$

whenever $X \leq Y$ in $\underline{X}$; also when $\underline{X}$ is a category by adjoining a family of axioms

$$D_f\colon X \longrightarrow Y\,,$$

one for each map $f\colon X \longrightarrow Y$ is $\underline{X}$. For technical reasons it is important to regard D_f and D_g as distinct axioms when $f \neq g$, even though they assert the same formula.

We shall lean heavily on concepts and techniques developed by logicians. However one question they seem to have ignored; namely, when are two proofs of the same formula equivalent? For example, consider the proof:

$$6\;\dfrac{4\;\dfrac{A \wedge B \xrightarrow{\;1\;} A \wedge B}{A \wedge B \longrightarrow A} \qquad\qquad 5\;\dfrac{A \wedge B \xrightarrow{\;1\;} A \wedge B}{A \wedge B \longrightarrow B}}{A \wedge B \longrightarrow A \wedge B}\,.$$

Nobody would deny that this proof might be regarded as, in some sense, equivalent to the shorter proof:

$$A \wedge B \xrightarrow{\;1\;} A \wedge B\,.$$

On the other hand, there is a good case for considering the following proofs as inequivalent:

$$4\;\dfrac{A \wedge A \xrightarrow{\;1\;} A \wedge A}{A \wedge A \longrightarrow A}\,, \qquad 5\;\dfrac{A \wedge A \xrightarrow{\;1\;} A \wedge A}{A \wedge A \longrightarrow A}\,.$$

In fact, they admit different "generalizations".

Our program will be to look at two kinds of structured categories in some detail: standard constructions and closed categories. It had been planned to discuss also cartesian closed categories and combinators, but time did not permit inclusion in this volume. A treatment of this topic and also the development of a general theory of structured categories must be deferred.

1. STANDARD CONSTRUCTIONS

By a <u>closure system</u> we shall understand a pair $(\underline{A},T)$ where $\underline{A}$ is a pre-ordered set with an operation T satisfying the following rules:

$$A \leq T(A) \ ,$$
$$T(T(A)) \leq T(A) \ ,$$
$$A \leq B \longrightarrow T(A) \leq T(B) \ ,$$

for all elements A and B of $\underline{A}$. It will be convenient to write $T^2(A)$ for $T(T(A))$, etc., also $T^0(A)$ for A.

As a device for constructing free closure systems it is useful to look at a deductive system $D(\underline{X})$ generated by a pre-ordered set $\underline{X}$. Its <u>terms</u> are elements of $\underline{X}$ and expressions of the form $T(A)$ when A is a term. Of course, instead of describing the set of terms recursively, we can, in this case, list all terms explicitely as being expressions of the form $T^n(X)$, where X is an element of

$\underline{X}$ and n is a non-negative integer. The formulas of $D(\underline{X})$ are expressions of the form $A \longrightarrow B$ where A and B are terms. They also may be listed explicitely as being expressions of the form $T^m(X) \longrightarrow T^n(Y)$. It remains to present the <u>axioms</u> and <u>rules of inference</u>:

$$A \xrightarrow{1_A} A \,,$$

$$\frac{A \xrightarrow{Q} B \qquad B \xrightarrow{P} C}{A \xrightarrow{PQ} C} \,,$$

$$X \xrightarrow{D_{X,Y}} Y \,, \text{ whenever } X \leq Y \text{ in } \underline{X},$$

$$A \xrightarrow{h(A)} T(A) \,, \qquad T^2(A) \xrightarrow{m(A)} T(A) \,,$$

$$\frac{A \xrightarrow{P} B}{T(A) \xrightarrow{T(P)} T(B)} \,.$$

The last rule of inference, for example, must be interpreted as a rule for generating a proof $T(P): T(A) \longrightarrow T(B)$ from a given proof $P: A \longrightarrow B$.

An example of a proof in $D(\underline{X})$ is the following:

$$\cfrac{\cfrac{\cfrac{A \xrightarrow{1_A} A}{T(A) \longrightarrow T(A)}}{T^2(A) \longrightarrow T^2(A)} \qquad T^2(A) \xrightarrow{m(A)} T(A)}{T^2(A) \longrightarrow T(A)} \,,$$

denoted unambiguously by $m(A)T^2(1_A)$.

We now construct the free closure system $F(\underline{X})$ generated by the pre-ordered set $\underline{X}$. Its elements are the terms of $D(\underline{X})$, and $A \leq B$ in $F(\underline{X})$ means that $A \longrightarrow B$ is provable in $D(\underline{X})$. $T(A)$ is the same in $F(\underline{X})$ as in $D(\underline{X})$. It is easily seen that $F(\underline{X})$ is a closure system. In what sense is it free?

By Ord we shall mean the category of pre-ordered sets and order preserving mappings. By Clo we shall mean the category of closure systems and those order preserving mappings which preserve T exactly, let us call them strict maps. Thus a strict map $H: (\underline{A},T) \longrightarrow (\underline{B},T)$ is a map $H: \underline{A} \longrightarrow \underline{B}$ in Ord such that

$$H(T(A)) = T(H(A))$$

for all elements A of $\underline{A}$. It is a sore point of our theory that "=" in the above cannot be replaced by "$\leq$ and $\geq$", otherwise we could not make the following statements without first generalizing the concept of "adjoint". There is an obvious forgetful functor U: Clo $\longrightarrow$ Ord. As is well known, a left adjoint F of U is determined by a solution to the universal mapping problem: Given any pre-ordered set $\underline{X}$, find a closure system $F(\underline{X})$ and an order preserving mapping $\delta(\underline{X}): \underline{X} \longrightarrow U(F(\underline{X}))$ so that for every closure system $(\underline{A},T)$ and every order preserving mapping $H: \underline{X} \longrightarrow \underline{A}$ there exists a unique strict map $H': F(\underline{X}) \longrightarrow \underline{A}$ for which

$U(H')\,\delta(X) = H.$

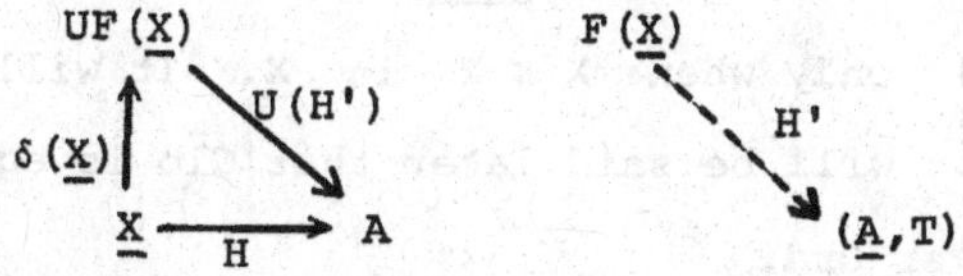

It remains to define $\delta(\underline{X})$. We stipulate that $\delta(\underline{X})(X) = X$ for each element X of $\underline{X}$ and verify that $\delta(\underline{X})$ preserves order. Indeed, if $X \leq Y$ in $\underline{X}$, then $X \longrightarrow Y$ is an axiom $D_{X,Y}$ in $D(\underline{X})$, hence provable and, therefore, $X \leq Y$ in $F(\underline{X})$.

Now, assume that $(\underline{A},T)$ is any closure system and $H\colon \underline{X} \longrightarrow \underline{A}$ any order preserving mapping. Define $H'\colon F(\underline{X}) \longrightarrow (\underline{A},T)$ recursively by $H'(X) = X$ and $H'(T(A)) = T(H'(A))$, or all at once by $H'(T^n(X)) = T^n(H(X))$. It is easily verified that H' is a strict map in Clo, in fact the only strict map for which $U(H')\,\delta(\underline{X}) = H$.

Here it should be pointed out that it is in fact easy to list all theorems of $D(\underline{X})$. They are precisely the formulas $T^m(X) \longrightarrow T^n(Y)$ where $X \leq Y$ in $\underline{X}$ and $m = 0$ or $m = n$ or $n > 1$. Thus $F(\underline{X})$ is isomorphic to the closure system whose elements are the non-negative integers, with the pre-order relation $m \leq n$ whenever $m = 0$ or $m = n$ or $n > 1$, and for which T is the successor function.

We shall see later, in the context of standard constructions, that $\delta(\underline{X})$ is full in the sense that $X \leq Y$ in $F(\underline{X})$ only when $X \leq Y$ in $\underline{X}$. It will also follow from what will be said later that Clo is equational (tripleable) over Ord.

A standard construction is a quadruple $(\underline{A},T,\eta,\mu)$ where $\underline{A}$ is a category, $T: \underline{A} \longrightarrow \underline{A}$ is a functor, and $\eta: 1_{\underline{A}} \longrightarrow T$ and $\mu: T^2 \longrightarrow T$ are natural transformations satisfying $\mu\cdot(\eta T) = 1_T = \mu\cdot(T\eta)$ and $\mu\cdot(\mu T) = \mu\cdot(T\mu)$. By Trip we shall mean the category whose objects are small standard constructions and whose maps, called strict maps, say from $(\underline{A},T,\eta,\mu)$ to $(\underline{B},T,\eta,\mu)$, are those functors $H: \underline{A} \longrightarrow \underline{B}$ which preserve the structure exactly, in the sense that

$$HT = TH, \; H\eta = \eta H, \; H\mu = \mu H .$$

What has been said in apology for using strict maps in Clo also applies here.

We wish to construct a left adjoint to the obvious forgetful functor $U: \text{Trip} \longrightarrow \text{Cat}$. To this purpose we slightly modify the deductive system $D(\underline{X})$ used earlier. Now $\underline{X}$ may be any small category, not just a pre-ordered set, and we must speak of the objects rather than of the elements of $\underline{X}$.

Also the axiom $D_{X,Y}$ is replaced by a family of axioms

$$D_f : X \longrightarrow Y ,$$

one for each map $f: X \longrightarrow Y$ in $\underline{X}$. It is understood that $D_f \neq D_g$ whenever $f \neq g$.

We now construct the free standard construction $F(\underline{X})$ together with a functor $\delta(\underline{X}): \underline{X} \longrightarrow U(F(\underline{X}))$. The objects of $F(\underline{X})$ are the terms of $D(\underline{X})$. The maps of $F(\underline{X})$ are equivalence classes of proofs of $D(\underline{X})$, according to a certain equivalence relation $\equiv$ between proofs, to be defined presently. We write $[P]$ for the set of all proofs $Q \equiv P: A \longrightarrow B$. Identity maps, composition of maps, T, η, μ, and $\delta(\underline{X})$ are defined thus:

$$1_A = [1_A],\ [P][Q] = [PQ],\ T([P]) = [T(P)],$$
$$\eta(A) = [h(A)],\ \mu(A) = [m(A)],$$
$$\delta(\underline{X})(X) = X,\ \delta(\underline{X})(f) = [D_f].$$

It remains to specify the equivalence relation $\equiv$. We take it to be the smallest transitive, symmetric, and reflexive relation on the set of proofs of $D(\underline{X})$ which satisfies the following rules:

(1) The substitution rules

$$\frac{P \equiv Q \quad R \equiv S}{PR \equiv QS}\ ,\quad \frac{P \equiv Q}{T(P) \equiv T(Q)}\ ,$$

because only by virtue of these rules are the above

definitions of [P][Q] and T([P]) justified.

(2) Rules which allow us to conclude that $UF(\underline{X})$ is a category:

$$1_B P \equiv P, \quad P 1_A \equiv P, \quad (PQ)R \equiv P(QR) .$$

The last rule, for example, asserts the equivalence of the following two proofs:

$$\frac{A \xrightarrow{R} B \qquad \dfrac{B \xrightarrow{Q} C \quad C \xrightarrow{P} A}{B \longrightarrow D}}{A \longrightarrow D}\ , \qquad \frac{\dfrac{A \xrightarrow{R} B \quad B \xrightarrow{Q} C}{A \longrightarrow C} \qquad C \xrightarrow{P} D}{A \longrightarrow D}\ .$$

(3) Rules which assure that T is a functor:

$$T(1_A) \equiv 1_{T(A)}, \ T(PQ) \equiv T(P)T(Q) .$$

(4) Rules which assure that η and μ are natural:

$$T(P)h(A) \equiv h(B)P, \ T(P)m(A) \equiv m(B)T^2(P) .$$

(5) Rules which assure the commutativity conditions for η and μ:

$$m(A)h(T(A)) \equiv 1_{T(A)} \equiv m(A)T(h(A)) ,$$

$$m(A)T(m(A)) \equiv m(A)m(T(A)) .$$

(6) Rules which assure that $\delta(\underline{X})$ is a functor:

$$D_{1_X} \equiv 1_X, \ D_{fg} \equiv D_f D_g .$$

PROPOSITION 1. (F,δ) is a solution of the universal mapping problem for U; that is, for each small category $\underline{X}$, each small standard constructions $(\underline{A},T,\eta,\mu)$, and each functor $H: \underline{X} \longrightarrow \underline{A}$, there exists a unique strict map $H': F(\underline{X}) \longrightarrow (\underline{A},T,\eta,\mu)$ such that $U(H')\ \delta(\underline{X}) = H$.

Proof. Given $H: \underline{X} \longrightarrow \underline{A}$, we construct H' as follows:

$$H'(X) = H(X),\ H'(T(A)) = T(H'(A)) \ .$$

Before defining $H'([P])$, we define $H'(P)$ for each proof P in $D(\underline{X})$ recursively thus:

$$H'(1_A) = 1_{H'(A)},\ H'(PQ) = H'(P)\ H'(Q),\ H'(T(P)) = T(H'(P)),$$

$$H'(h(A)) = \eta(H'(A)),\ H'(m(A)) = \mu(H'(A)),\ H'(D_f) = H(f) \ .$$

It is easily verified that

$$P \equiv Q \Longrightarrow H'(P) = H'(Q) \ ,$$

by induction on the length of the proof of equivalence. Consequently, we are justified in defining

$$H'([P]) = H'(P) \ .$$

Moreover, one quickly verifies that

$$H'(T(A)) = T(H'(A)),\ H'(T([P]) = T(H'([P]),$$
$$H'(\eta(A)) = \eta(H'(A)),\quad H'(\mu(A)) = \mu(H'(A)),$$

hence H' is indeed a strict map in Trip.

Finally

$$U(H')\ \delta(\underline{X})(X) = H(X),\ U(H')\ \delta(\underline{X})(f) = H(f)\ ,$$

and these equations force the definition of H'.

REMARK 1. As is well known, F can be made into a functor and δ into a natural transformation such that F becomes the left adjoint of U with adjunction $\delta: 1 \longrightarrow UF$. In the present situation one can easily extend the isomorphism $[F(\underline{X}),\ (\underline{A},T,\eta,\mu)] \cong [\underline{X},\underline{A}]$ in Ens to one in Cat. However this is not typical for structured categories and depends on the fact that T is a covariant functor.

REMARK 2. It may easily be shown at this stage that Trip is equational (tripleable) over Cat by verifying Beck's conditions. However the verification is too easy in this example to be typical, and we shall defer it to another structured category.

PROPOSITION 2. $\delta(\underline{X})$ is full and faithful.

Proof. (a) Given a proof $P: X \longrightarrow Y$ in $D(\underline{X})$, we claim that $P \equiv D_f$ for some map $f: X \longrightarrow Y$ in $\underline{X}$. Actually we shall show more generally that any proof $P: A \longrightarrow Y$ has this form. We proceed by induction on the length of P.

If $P = 1_Y$, then surely $P \equiv D_{1_Y}$. If $P = D_f$, there is nothing to prove. $P = T(A)$, $h(A)$, or $m(A)$ are all ruled out. There only remains $P = QR$, where $R: A \longrightarrow B$ and $Q: B \longrightarrow Y$. By inductional assumption we have $Q \equiv D_f$ for some $f: Z \longrightarrow Y$. Hence, again by inductional assumption, $R \equiv D_g$ for some $g: X \longrightarrow Z$. Therefore $P \equiv D_f D_g \equiv D_{fg}$.

An easier proof of (a) will be given later.

(b) Suppose $D_f \equiv D_g$, we wish to prove that $f = g$. We need a definition and a lemma. $\underline{2}$ is the category with two objects 0 and 1, and one non-identity map $0 \longrightarrow 1$.

DEFINITION. The <u>generality</u> of a proof P in $D(\underline{X})$ is the set of all functors $H: \underline{2} \longrightarrow \underline{X}$ for which there exists a proof P' in $D(\underline{2})$ such that $P = H(P')$. By the last equation we mean that P results from P' if all occurrences of 0, 1, and $\longrightarrow$ in P' are replaced by $H(0)$, $H(1)$, and $H(\longrightarrow)$ respectively.

LEMMA 1. Equivalent proofs have the same generality.

Assuming this for the moment, we shall continue with (b). Let $H: \underline{2} \longrightarrow \underline{X}$ be defined by

$$H(0) = X, H(1) = Y, H(\longrightarrow) = f \ .$$

Then $H(D_{\longrightarrow}) = D_f$. Therefore, by the lemma, there exists a map $x: 0 \longrightarrow 1$ in $\underline{2}$ such that $H(D_x) = D_g$. Clearly, $x = \longrightarrow$, hence $f = g$. It remains to prove the lemma.

Proof sketched. We proceed by induction on the length of the proof that the given proofs are equivalent. For example, assume that the last step in the proof of equivalence was a substitution rule, say

$$\frac{P \equiv Q}{T(P) \equiv T(Q)}$$

By inductional assumption, P and Q have the same generality. Let $H: \underline{2} \longrightarrow \underline{X}$ and $H(P^*) = P$, where $P^*: T(A') \longrightarrow T(B')$ in $D(\underline{2})$. Then $P^* = T(P')$, where $P': A' \longrightarrow B'$ and $H(P') = P$. Therefore, there exists $Q': A' \longrightarrow B'$ with $H(Q') = Q$, hence $H(T(Q')) = T(H(Q')) = T(Q)$.

We hasten to point out that the converse of the above lemma does not hold. For example, the proof

$$\frac{T^2(A) \xrightarrow{m(A)} T(A) \qquad T(A) \xrightarrow{h(T(A))} T^2(A)}{T^2(A) \longrightarrow T^2(A)}$$

is not equivalent to the axiom $T^2(A) \longrightarrow T^2(A)$, although it is easily seen to have the same generality. Indeed, supposing

$$h(T(A))\, m(A) \equiv 1_{T^2(A)} \qquad \ldots\ldots (7)$$

we should have

$$\eta(T(A))\ \mu(A) = 1_{T^2(A)}$$

in $F(\underline{X})$. But then it would easily follow that $T^2(A) \cong T(A)$ not only in $F(\underline{X})$, but in every small standard construction. To see that this is not so, take for example $\underline{A}$ = the category of countable sets, $T(A)$ = the underlying set of the free group generated by A, etc.

A standard construction is called <u>idempotent</u> if $\eta T \cdot \mu = 1_{T^2}$. For idempotent constructions we have the same deductive system as for standard constructions, but one additional rule to be satisfied by the equivalence relation on proofs. We shall see later that the converse of the above lemma does hold for idempotent constructions.

We are interested in obtaining a decision procedure for the deductive system $D(\underline{X})$: Given terms A and B, can we find all proofs of the formula $A \longrightarrow B$ at least up to equivalence? Of course this presumes that all maps $f: X \longrightarrow Y$ in $\underline{X}$ are known. Note that a decision procedure for the mere existence of proofs is easy: $T^m(X) \longrightarrow T^n(Y)$ is a theorem if and only if there exists a map $f: X \longrightarrow Y$ and $m = 0$ or $m = n$ or $n > 1$.

First we shall replace $D(\underline{X})$ by a Gentzen-type deductive system $G(\underline{X})$ with the same terms and formulas.

However $G(\underline{X})$ has the following axioms and rules of inference:

$$D_f : X \longrightarrow Y \quad \text{when} \quad f : X \longrightarrow Y \quad \text{in} \quad \underline{X} ,$$

$$\frac{A \xrightarrow{P} B}{A \xrightarrow{r(P)} T(B)} , \qquad \frac{A \xrightarrow{P} T(B)}{T(A) \xrightarrow{l(P)} T(B)} .$$

These are derived rules in $D(\underline{X})$. In fact the proofs

$$\frac{A \longrightarrow B \quad B \longrightarrow T(B)}{A \longrightarrow T(B)} , \qquad \frac{\dfrac{A \longrightarrow T(B)}{T(A) \longrightarrow T^2(B)} \quad T^2(B) \longrightarrow T(B)}{T(A) \longrightarrow T(B)}$$

show that we may define

$$r(P) = h(B)P , \quad l(P) = m(B)T(P) .$$

In view of these definitions, each proof of $G(\underline{X})$ may be expanded to a uniquely determined proof of $D(\underline{X})$. In the converse direction we have the following.

PROPOSITION 3. Each proof in $D(\underline{X})$ is equivalent to (the expansion of) a proof in $G(\underline{X})$.

Proof. For any proof $P : A \longrightarrow B$ in $G(\underline{X})$, we have

$$T(P) \equiv 1_{T(B)} T(P) \equiv m(B)T(h(B))T(P) \equiv m(B)T(h(B)P) = l(r(P)) .$$

Next we shall prove by induction on the number of occurrences of T in A that 1_A is equivalent to a proof in $G(\underline{X})$.

Surely $1_X \equiv D_{1_X}$. Now assume the result for A, say $1_A \equiv P_A$ in $G(\underline{X})$, then

$$1_{T(A)} \equiv T(1_A) \equiv T(P_A) \equiv l(r(P_A))$$

by the above. Again, we have

$$h(A) \equiv h(A)1_A = r(1_A) \equiv r(P_A) \;,$$

$$m(A) \equiv m(A)1_{T^2(A)} \equiv m(A)T(1_{T(A)}) = l(1_{T(A)}) \equiv l(P_{T(A)}) \;.$$

There only remains transitivity. This is handled by the following:

CUT ELIMINATION THEOREM. If $Q\colon A \longrightarrow B$ and $P\colon B \longrightarrow C$ are proofs in $G(\underline{X})$, then PQ is equivalent to a proof in $G(\underline{X})$.

Proof. We proceed by induction on the number of occurrences of T in A, B, C.

Case 1. Both premises are instances of D_f. Then $D_f D_g \equiv D_{fg}$.

Case 2. The last step in the proof of the second premise uses r. Then we have

$$\frac{A \xrightarrow{Q} B \qquad \dfrac{B \xrightarrow{R} C}{B \longrightarrow T(C)}}{A \longrightarrow T(C)} \;\equiv\; \frac{\dfrac{A \xrightarrow{Q} B \qquad B \xrightarrow{R} C}{A \longrightarrow C}}{A \longrightarrow T(C)}$$

since

$$r(R)Q = (h(C)R)Q \equiv h(C)(RQ) = r(RQ) \ .$$

By inductional assumption RQ is equivalent to some proof in $G(\underline{X})$.

Case 3. The last steps in the proofs of both premises use 1. Then we have

$$\frac{\dfrac{A \xrightarrow{S} T(B)}{T(A) \longrightarrow T(B)} \quad \dfrac{B \xrightarrow{R} T(C)}{T(B) \longrightarrow T(C)}}{T(A) \longrightarrow T(C)} \equiv \dfrac{\dfrac{A \xrightarrow{S} T(B) \qquad \dfrac{B \xrightarrow{R} T(C)}{T(B) \longrightarrow T(C)}}{A \longrightarrow T(C)}}{T(A) \longrightarrow T(C)}$$

since

$$\begin{aligned} 1(R)1(S) &= m(D)T(R)\ m(B)T(S) \\ &\equiv m(D)\ m(T(D))T^2(R)T(S) \\ &\equiv m(D)\ T(m(D)T(R)S) \\ &= 1(1(R)S) \ . \end{aligned}$$

By inductional assumption $1(R)S$ is equivalent to some proof in $G(\underline{X})$.

Case 4. The last step in the proof of the first premise uses r and the last step in the proof of the second premise uses 1. Then we have

$$\frac{\dfrac{A \xrightarrow{S} B}{A \longrightarrow T(B)} \quad \dfrac{B \xrightarrow{R} T(C)}{T(B) \longrightarrow T(C)}}{A \longrightarrow T(C)} \equiv \frac{A \xrightarrow{S} B \quad B \xrightarrow{R} T(C)}{A \longrightarrow T(C)}$$

since

$$1(R)r(S) = m(C)T(R)h(B)S \equiv m(C)h(T(C)RS \equiv RS .$$

By inductional assumption RS is equivalent to a proof in $G(\underline{X})$.

The proof of the theorem is now complete.

The advantage of the system $G(\underline{X})$ over the system $D(\underline{X})$ is this: given a formula $A \longrightarrow B$, we can find all possible proofs of this formula in a finite number of steps, by working backwards.

To give one application let us present a new proof of Proposition 2(b); that is, that $\delta(\underline{X})$ is full. Assume that $P: X \longrightarrow Y$ in $D(\underline{X})$. By Proposition 3, P is equivalent to some proof Q in $G(\underline{X})$. Since T does not occur in the formula $X \longrightarrow Y$, Q must be D_f for some $f: X \longrightarrow Y$.

As another application, one may translate the recursive description of the proofs in $G(\underline{X})$ into a recursive description of the maps in $F(\underline{X})$. It then easily follows that $F(\underline{X}) = F(\underline{1}) \times X$. This result is due to Bill Lawvere, who also observed that $F(\underline{1})$ is isomorphic with the category of finite ordinals and order preserving maps.

We are now in a position to obtain the converse of Lemma 1 for idempotent constructions.

COHERENCE THEOREM FOR IDEMPOTENT CONSTRUCTIONS. Two proofs in $D(\underline{X})$ with the same generality are equivalent.

Proof. Since each proof in $D(\underline{X})$ is equivalent to one in $G(\underline{X})$ and therefore has the same generality, we need only consider proofs in $G(\underline{X})$.

Let $P: A \longrightarrow B$ be a proof in $G(\underline{X})$. It contains exactly one axiom $D_f: X \longrightarrow Y$. Define $H: \underline{2} \longrightarrow \underline{X}$ by

$$H(0) = X, \; H(1) = Y, \; H(\longrightarrow) = f \; .$$

Let P' be the proof in $G(\underline{2})$ obtained from P by replacing f by $\longrightarrow$. Then $H(P') = P$. Now, assume that $Q: A \longrightarrow B$ has the same generality as P. Then there exists a proof Q^* in $D(\underline{2})$ such that $H(Q^*) = Q$. But, by Proposition 3, Q^* is equivalent to some proof Q' in $G(\underline{2})$, hence, also $H(Q^*) \equiv H(Q')$. It follows from the lemma below that $P' \equiv Q'$, hence

$$P = H(P') \equiv H(Q') \equiv H(Q^*) = Q \; .$$

LEMMA 2. For idempotent constructions, any two proofs of $T^m(0) \longrightarrow T^n(1)$ in $G(\underline{2})$ are equivalent.

Proof. We proceed by induction on $m + n$. Aside from symmetry, we need only consider the following cases:

$$(P,Q) = (D_f,D_g), \; (l(P'),l(Q')), \; (r(P'),r(Q')), \; (l(P'),r(Q')) \; .$$

In the first case f and g are maps $0 \longrightarrow 1$, hence $f = g$.

In the second case we have $P' \equiv Q'$ by inductional assumption, hence $P \equiv Q$.

The third case is similar to the second.

We shall consider the fourth case after the following:

SUBLEMMA. If $T(A) \longrightarrow B$ is provable in $G(\underline{X})$ then $B = T(C)$ and $A \longrightarrow B$ is also provable.

Proof. This is shown by induction on the length of the proof of $T(A) \longrightarrow B$. If the last step in this proof is l, there is nothing to show, so we may assume it is r. Then $B = T(C)$ and $T(A) \longrightarrow C$ is a theorem. By inductional assumption, $C = T(D)$ and $A \longrightarrow C$ is provable, hence so is $A \longrightarrow T(C) = B$.

We now return to the fourth case in the proof of Lemma 2.

We have proofs $P = l(P')$ and $Q = r(Q')$ of $T(A) \longrightarrow T(B)$ in $G(\underline{2})$, where $P': A \longrightarrow T(B)$ and $Q': T(A) \longrightarrow B$. In view of the sublemma, $B = T(C)$ and there is a proof $R: A \longrightarrow B$. By inductional assumption, $P' \equiv r(R)$ and $Q' \equiv l(R)$, hence,

$$P \equiv l(r(R)) = m(B)T(T(R)h(A))$$
$$\equiv m(B)T^2(R)T(h(A))$$
$$\equiv T(R)m(A)T(h(A)) \equiv T(R)$$

and

$$Q \equiv r(l(R)) = h(B)m(C)T(R)$$
$$= h(T(C))m(C)T(R) \equiv T(R)$$

by the idempotent rule (7). This is the only place where the idempotent rule comes in.

2. CLOSED CATEGORIES

We might as well start with a bang and define a <u>biclosed</u> <u>monoidal</u> category as a 10-tuple $(\underline{A}, I, \cdot, /, \backslash, \rho, \lambda, \alpha, \beta, \gamma)$ where

$\underline{A}$ is a category,

I is an object of $\underline{A}$,

$\cdot : \underline{A} \times \underline{A} \longrightarrow \underline{A}$, $/ : \underline{A} \times \underline{A}^{opp} \longrightarrow \underline{A}$, and $\backslash : \underline{A}^{opp} \times \underline{A} \longrightarrow \underline{A}$ are bifunctors,

$\rho, \lambda, \alpha, \beta, \gamma$ are natural isomorphisms:

$$\rho(A) : A \cdot I \longrightarrow A, \quad \lambda(A) : I \cdot A \longrightarrow A ,$$
$$\alpha(A,B,C) : (A \cdot B) \cdot C \longrightarrow A \cdot (B \cdot C) ,$$
$$\beta(A,B,C) : [A \cdot B, C] \longrightarrow [A, C/B] ,$$
$$\gamma(A,B,C) : [A \cdot B, C] \longrightarrow [B, A \backslash C] ,$$

such that the following composite maps are identities:

$$A\cdot B \longrightarrow (A\cdot I)\cdot B \longrightarrow A\cdot(I\cdot B) \longrightarrow A\cdot B \,,$$

$$((A\cdot B)\cdot C)\cdot D \longrightarrow (A\cdot B)\cdot(C\cdot D) \longrightarrow A\cdot(B\cdot(C\cdot D)) \longrightarrow$$
$$A\cdot((B\cdot C)\cdot D) \longrightarrow (A\cdot(B\cdot C))\cdot D \longrightarrow ((A\cdot B)\cdot C)\cdot D \,.$$

Some historical remarks are in order. Monoidal categories $(\underline{A}, I, \cdot, \rho, \lambda, \alpha)$ were studied by Benabou (1963). The last two "coherence" conditions were discussed by MacLane (1963). Closed monoidal categories $(\underline{A}, I, \cdot, \backslash, \rho, \lambda, \alpha, \gamma)$ were studied by Eilenberg and Kelly (1966). Residuated categories $(\underline{A}, \cdot, /, \backslash, \alpha, \beta, \gamma)$ were studied in Deductive Systems and Categories (I). However one cannot strip off $\cdot$ and α without making some other changes, as is clear from the paper by Eilenberg and Kelly. The deductive system corresponding to residuated categories had already been studied in 1958 with a view of applications to mathematical linguistics.

With minor changes the proofs in Deductive Systems and Categories (I) go through to establish the following results:

(1) A free biclosed monoidal category $F(\underline{X})$ together with a functor $\delta(\underline{X}) : \underline{X} \longrightarrow U(F(\underline{X}))$ can be constructed for any small category $\underline{X}$ using a suitable deductive system $D(\underline{X})$.

(2) $\delta(\underline{X})$ is a full embedding.

(3) $D(\underline{X})$ may be replaced by a Gentzen-type system $G(\underline{X})$, allowing one to find all proofs of a formula $A \longrightarrow B$ in $D(\underline{X})$ up to equivalence.

(4) Two proofs of $A \longrightarrow B$ in $D(\underline{X})$ are equivalent if and only if they have the same generality.

To explain the concept of generality (called "scope" in the paper referred to) let us consider two possible definitions.

DEFINITION A. The <u>Generality</u> of a proof $P: A \longrightarrow B$ in $D(\underline{X})$ is the class of all pairs $(\underline{X}', H)$ where $\underline{X}'$ is any small category and $H: \underline{X}' \longrightarrow \underline{X}$ is a functor such that there exists a proof P' in $D(\underline{X}')$ with $H(P') = P$.

Here $H(P')$ is the proof obtained from P by replacing each object and each map of $\underline{X}'$ occurring in P' by its image under H.

It turns out that when testing whether two proofs have the same Generality we need not look at all small categories $\underline{X}'$ but only at categories of the form $n\underline{2}$ (sum of n disjoint copies of $\underline{2}$), where n depends on the width of the given proofs. We make the following alternative definition.

DEFINITION B. The generality of a proof $P: A \longrightarrow B$ in $D(\underline{X})$ is the set of all pairs $(\underline{X}', H)$ where $\underline{X}' = n\underline{2}$ for some finite n and . . . . . .

We note that (3) and (4) together allow us to compute $[A,B]$ in $F(\underline{X})$, assuming of course that $[X,Y]$ in $\underline{X}$ is known.

A functor H between biclosed monoidal categories is called a strict map if it preserves the structure exactly, that is,

$$H(I) = I, H(\cdot) = H(\ )\cdot H(\ ), \text{ etc.}$$

$$H\rho = \rho H, \text{ etc}, H(\alpha(A,B,C)) = \alpha(H(A), H(B), H(C)),$$

$$H(\beta(A,B,C)(f)) = \beta(H(A), H(B), H(C))(H(f)), \text{ etc.}$$

Let BMC denote the category of small biclosed monoidal categories, then the forgetful functor $U: \text{BMC} \longrightarrow \text{Cat}$ has a left adjoint F. We may now ask whether BMC consists of exactly the algebras of the standard construction $(UF, \delta, -)$ in the sense of Beck's Precise Tripleableness Theorem. Beck's conditions, as modified by Paré, demand this:

(1) U reflects isomorphisms.

(2) Given two strict maps $H_1, H_2: (\underline{A}, \ldots) \longrightarrow (\underline{B}, \ldots)$ in BMC and a functor $K: \underline{B} \longrightarrow \underline{C}$ such that $\underline{A} \rightrightarrows \underline{B} \longrightarrow \underline{C}$ is an absolute coequalizer diagram (that is, $L(\underline{A}) \rightrightarrows L(\underline{B}) \longrightarrow L(\underline{C})$ is a coequalizer diagram for every functor L with domain Cat), then the pair H_1, H_2 has a

coequalizer K' and U(K') ≅ K canonically.

We shall not give here the complete proof that BMC does indeed satisfy the Beck-Paré conditions, but we shall look at a crucial part of the argument for condition (2).

Assume that H_1, H_2, and K are given and consider the following diagram:

$$\begin{array}{ccccc} \underline{A}^{opp} \times \underline{A} & \rightrightarrows & \underline{B}^{opp} \times \underline{B} & \longrightarrow & \underline{C}^{opp} \times \underline{C} \\ \backslash\downarrow & & \backslash\downarrow & & \downarrow \\ \underline{A} & \rightrightarrows & \underline{B} & \longrightarrow & \underline{C} \end{array}$$

Since H_1 and H_2 are strict maps, the two squares on the left commute. Since the bottom is an absolute coequalizer diagram, the top is a coequalizer diagram. It follows that there is a unique functor $\underline{C}^{opp} \times \underline{C} \longrightarrow \underline{C}$ to make the right square commute, let us also denote it by $\backslash$. Then K preserves $\backslash$ exactly.

We may deal similarly with the functors $\cdot$ and $/$, but what about the natural transformations ρ to γ? The trick is to view these also as functors, thus for example $\beta: \underline{A}^{opp} \times \underline{A}^{opp} \times \underline{A} \longrightarrow Ens^2$ is the functor which, when followed by the canonical functor $Ens^2 \longrightarrow Ens \times Ens$ assigning to each map $f: X \longrightarrow Y$ the pair (X,Y) yields the functor $\underline{A}^{opp} \times \underline{A}^{opp} \times \underline{A} \longrightarrow Ens \times Ens$ whose value at

(A,B,C) is ([A·B,C],[A,C/B]).

We shall now turn to another way of studying closed categories by viewing them as multicategories. The connection between Gentzen's method in logic and multilinear mappings in algebra has been pointed out before (1961). I am told that multicategories have also been studied by Benabou and Cartier.

A <u>multicategory</u> consists of a class of objects together with a class of <u>multimaps</u>

$$g: A_1, A_2, \ldots, A_n \longrightarrow B ,$$

n being any non-negative integer. Among the multimaps are the <u>identity</u> maps $1_A: A \longrightarrow A$. Multimaps may be composed by "substitution" as follows: Given multimaps

$$g: A_1, \ldots, A_n \longrightarrow B, \quad f: \ldots, A, \ldots \longrightarrow B ,$$

there is a multimap

$$f(\ldots,g,\ldots): \ldots, A_1, \ldots, A_n, \ldots \longrightarrow B .$$

Substitution, also called <u>cut</u>, must satisfy four conditions. These will be stated as soon as we have introduced a convenient notation.

We shall use capital Greek letters to stand for finite sequences of objects.

Thus a multimap may be denoted by $g\colon \Lambda \longrightarrow B$. We write $\emptyset$ for the empty sequence, for which $n = 0$. A multimap $b\colon \emptyset \longrightarrow B$ may also be regarded as an <u>element</u> of B. If Φ and Ψ are finite sequences of objects, their juxtaposition is denoted by Φ,Ψ. The reader should ignore the comma when Φ or Ψ is the empty sequence. The cut now takes the form

$$\frac{\Lambda \xrightarrow{g} A \quad \Phi,A,\Psi \xrightarrow{f} B}{\Phi,\Lambda,\Psi \longrightarrow B}.$$

We shall avoid the notation $f(\Phi,g,\Psi)$ for the resulting multimap as it tends to become troublesome. We now state the four conditions.

(1) The following multimap is equal to f:

$$\frac{A \xrightarrow{1_A} A \quad \Phi,A,\Psi \xrightarrow{f} B}{\Phi,A,\Psi \longrightarrow B}.$$

(2) The following multimap is equal to g:

$$\frac{\Lambda \xrightarrow{g} A \quad A \xrightarrow{1_A} A}{\Phi,\Lambda,\Psi \longrightarrow B}.$$

(3) The following multimaps are equal:

$$\frac{\dfrac{\Lambda \longrightarrow A \quad \Phi,A,\Psi \longrightarrow B}{\Phi,\Lambda,\Psi \longrightarrow B} \quad \Gamma,B,\Delta \longrightarrow C}{\Gamma,\Phi,\Lambda,\Psi,\Delta \longrightarrow C}, \qquad \frac{\Lambda \longrightarrow A \quad \dfrac{\Phi,A,\Psi \longrightarrow B \quad \Gamma,B,\Delta \longrightarrow C}{\Gamma,\Phi,A,\Psi,\Delta \longrightarrow B}}{\Gamma,\Phi,\Lambda,\Psi,\Delta \longrightarrow C}.$$

(4) The following multimaps are equal:

$$\frac{\Gamma \longrightarrow C \qquad \dfrac{\Delta \longrightarrow D \quad \Phi,C,\Theta,D,\Psi \longrightarrow B}{\Phi,C,\Theta,\Delta,\Psi \longrightarrow B}}{\Phi,\Gamma,\Theta,\Delta,\Psi \longrightarrow C}\ , \qquad \frac{\Delta \longrightarrow D \qquad \dfrac{\Gamma \longrightarrow C \quad \Phi,C,\Theta,D,\Psi \longrightarrow B}{\Phi,\Gamma,\Theta,D,\Psi \longrightarrow B}}{\Phi,\Gamma,\Theta,\Delta,\Psi \longrightarrow C}\ .$$

We may call (1) and (2) *identity* laws, (3) the *associative* law, and (4) the *commutative* law for multicategories.

We denote the class of all multimaps $\Lambda \longrightarrow B$ by $[\Lambda;B]$. The reader will appreciate the reason for use of the semicolon.

EXAMPLES

1. Any category becomes a multicat if we put $[A_1, \ldots, A_n; B] = \emptyset$ for $n \neq 1$.

2. Any monoidal category becomes a multicat if we define recursively

$$[\emptyset;B] = [I;B] \ ,$$

$$[\Gamma;B] = [C;B] \longrightarrow [\Gamma,A;B] = [C\cdot A;B] \ .$$

Thus, for example,

$$[A,B,C;D] = [(A\cdot B)\cdot C;D] \ .$$

3. Any (left) closed category becomes a multicat if we define recursively

$$[\emptyset;B] = [I;B], \ [C,\Gamma;B] = [\Gamma;C\backslash B] \ .$$

Thus, for example,

$$[A,B,C;D] = [C;B\backslash(A\backslash D)] = [I;C\backslash(B\backslash(A\backslash D))] .$$

We also mention one application. If $\underline{M}$ is any given multicategory, we may define $\underline{M}$-categories, that is categories for which $[A;B]$ is in some sense an object of $\underline{M}$. This may be done in the same way as it was done by Benabou (1965) for monoidal categories.

Certain structures are more easily introduced into multicats than into categories. Thus we may readily define monoidal multicats, left closed multicats, right closed multicats, and biclosed multicats by leaving out unwanted data from the following.

A <u>biclosed</u> <u>monoidal</u> <u>multicategory</u> is a 9-tuple $(\underline{M}, I, \cdot, /, \backslash, i, m, r, l)$ where $\underline{M}$ is a multicat, $\cdot$, $/$, and $\backslash$ are binary operations on the set of objects of $\underline{M}$, and $i: \emptyset \longrightarrow I$, $m: A,B \longrightarrow A\cdot B$, $r: A/B,B \longrightarrow A$, and $l: A,A\backslash B \longrightarrow B$ are multimaps such that the following induced mappings are one-one and onto:

$$[\Phi, I, \Psi; C] \longrightarrow [\Phi, \Psi; C] ,$$

$$[\Phi, A\cdot B, \Psi; C] \longrightarrow [\Phi, A, B, \Psi; C] ,$$

$$[\Gamma; A/B] \longrightarrow [\Gamma, B; A] ,$$

$$[\Gamma; B\backslash A] \longrightarrow [B, \Gamma; A] .$$

The following cuts show how, for example, the first and third of the above mappings are induced:

$$\frac{\emptyset \xrightarrow{i} I \qquad \Phi, I, \Psi \longrightarrow C}{\Phi, \Psi \longrightarrow C} \ , \ \frac{\Gamma \longrightarrow A/B \qquad A/B, B \xrightarrow{r} A}{\Gamma, B \longrightarrow A} \ .$$

A <u>functor</u> $E: \underline{M} \longrightarrow \underline{N}$ between multicategories associates with each object A of $\underline{M}$ an object $E(A)$ of $\underline{N}$ and with each multimap $f: A_1, \ldots, A_n \longrightarrow B$ a multimap $E(f): E(A_1), \ldots, E(A_n) \longrightarrow B$, and it preserves identities and cuts. Let Mult be the category of small multicategories and functors.

By a <u>strict</u> map E between biclosed monoidal multicategories, we mean a functor which preserves the structure exactly. Thus

$E(I) = I$, $E(A \cdot B) = E(A) \cdot E(B)$, etc., $E(i) = i$,

$E(m) = m$, etc.

Let BMM be the category of small biclosed monoidal multicategories and strict maps. There are obvious functors BMM $\longrightarrow$ BMC and BMC $\longrightarrow$ BMM. They establish an equivalence of categories, but it is rather tedious to show this in detail. Here is a description of these functors, at least on objects.

BMM $\longrightarrow$ BMC. Given any biclosed monoidal multicategory $(\underline{M}, I, \cdot, /, \backslash, i, m, r, l)$, form $(\underline{A}, I, \cdot, /, \backslash, \rho, \lambda, \alpha, \beta, \gamma)$ as follows. Suppress all multimaps which are not maps and define $\rho, \alpha, \beta, \beta^{-1}$, for example, as follows:

$$\dfrac{\dfrac{A \longrightarrow A}{A,I \longrightarrow A}}{A\cdot I \longrightarrow A}\ , \qquad \dfrac{\dfrac{\dfrac{B,C \longrightarrow B\cdot C \quad A, B\cdot C \longrightarrow A\cdot(B\cdot C)}{A,B,C \longrightarrow A\cdot(B\cdot C)}}{A\cdot B,C \longrightarrow A\cdot(B\cdot C)}}{(A\cdot B)\cdot C \longrightarrow A\cdot(B\cdot C)}\ ,$$

$$\dfrac{\dfrac{A,B \longrightarrow A\cdot B \quad A\cdot B \longrightarrow C}{A,B \longrightarrow C}}{A \longrightarrow C/B}\ , \qquad \dfrac{\dfrac{A \longrightarrow C/B \quad C/B,B \longrightarrow C}{A,B \longrightarrow C}}{A\cdot B \longrightarrow C}\ .$$

BMC $\longrightarrow$ BMM. Given any biclosed monoidal category $(\underline{A}, I, \cdot, /, \backslash, \rho, \lambda, \alpha, \beta, \gamma)$, form $(\underline{M}, I, \cdot, /, \backslash, m, r, l)$ as follows. Define $A_1, \ldots, A_n \longrightarrow B$ as in Example 2 above. Let

$$i\colon \emptyset \longrightarrow I \ \text{ mean } \ 1_I\colon I \longrightarrow I\ ,$$

$$m\colon A,B \longrightarrow A\cdot B \ \text{ mean } \ 1_{A\cdot B}\colon A\cdot B \longrightarrow A\cdot B\ ,$$

$$r\colon A/B,\ B \longrightarrow A \ \text{ mean } \ \beta^{-1}(1_{A/B})\colon A/B \cdot B \longrightarrow A$$

and similarly for l.

We shall now turn our attention to BMM. Given any small multicategory $\underline{X}$, we can form a deductive system $M(\underline{X})$ whose terms are the objects of $\underline{X}$; I, and $A\cdot B$, A/B, $A\backslash B$ if A and B are terms; and whose formulas have the form

$A_1, A_2, \ldots, A_n \longrightarrow B$ where the A_i and B are terms. Its axioms and rules of inference are as follows:

$$1_A : A \longrightarrow A\ ; \quad \text{the cut;} \quad M_f : X_1, \ldots, X_n \longrightarrow Y\ ,$$

$$\text{whenever} \quad f: X, \ldots, X_n \longrightarrow Y \text{ in } \underline{X}\ ;$$

$$\emptyset \longrightarrow I;\ A,B \longrightarrow A\cdot B;\ A/B,B \longrightarrow A;\ A,\ A\backslash B \longrightarrow B\ ;$$

$$\frac{\Phi,\Psi \longrightarrow C}{\Phi,I,\Psi \longrightarrow C}\ ; \quad \frac{\Phi,A,B,\Psi \longrightarrow C}{\Phi,A\cdot B,\Psi \longrightarrow C}\ ; \quad \frac{\Lambda,B \longrightarrow A}{\Lambda \longrightarrow A/B}\ ; \quad \frac{B,\Lambda \longrightarrow A}{\Lambda \longrightarrow B\backslash A}\ .$$

It is clear that we can introduce a suitable equivalence relation on $M(\underline{X})$ and use it to construct the free biclosed monoidal multicategory $F(\underline{X})$ generated by $\underline{X}$ and a functor $\delta(\underline{X}): \underline{X} \longrightarrow U(F(\underline{X}))$ showing that F is the left adjoint of U. Again, BMM will be equational over Mult.

We shall now replace $M(\underline{X})$ by a Gentzen-type deductive system $G(\underline{X})$. Its terms and formulas are the same as those of $M(\underline{X})$. Here are its axioms and rules of inference:

$$M_f : X_1, \ldots, X_n \longrightarrow Y$$

$$\text{whenever} \quad f: X_1, \ldots, X_n \longrightarrow Y \text{ in } \underline{X}\ ,$$

$\longrightarrow I$ $\qquad \emptyset \longrightarrow I$ $\qquad \dfrac{\Phi,\Psi \longrightarrow C}{\Phi,I,\Psi \longrightarrow C}$ $\qquad I \longrightarrow$

$\longrightarrow \cdot$ $\qquad \dfrac{\Gamma \longrightarrow C \qquad \Delta \longrightarrow D}{\Gamma,\Delta \longrightarrow C\cdot D}$ $\qquad \dfrac{\Phi,A,B,\Psi \longrightarrow C}{\Phi,A\cdot B,\Psi \longrightarrow C}$ $\qquad \cdot \longrightarrow$

$$\longrightarrow / \qquad \frac{\Lambda, B \longrightarrow A}{\Lambda \longrightarrow A/B} \qquad \frac{\Lambda \longrightarrow B \quad \Phi, A, \Psi \longrightarrow C}{\Phi, A/B, \Lambda, \Psi \longrightarrow C} \qquad / \longrightarrow$$

$$\longrightarrow \backslash \qquad \frac{B, \Lambda \longrightarrow A}{\Lambda \longrightarrow B\backslash A} \qquad \frac{\Lambda \longrightarrow B \quad \Phi, A, \Psi \longrightarrow C}{\Phi, \Lambda, B\backslash A, \Psi \longrightarrow C} \qquad \backslash \longrightarrow$$

Note that, aside from the axiom M_f which involves none of the symbols $\cdot$, $/$, $\backslash$, half the rules serve to introduce one of these symbols on the right and the other half serve to introduce one on the left. The notation $\longrightarrow /$ and $/ \longrightarrow$ is found in Kleene's book. Surely, given a formula in $G(\underline{X})$, we can find all possible proofs of it by working backwards, eliminating one symbol at each step.

PROPOSITION 1. Every rule of $G(\underline{X})$ is a derived rule in $M(\underline{X})$.

Proof. The only new rules are $\longrightarrow \cdot$, $/ \longrightarrow$, and $\backslash \longrightarrow$. Here is how the first two are shown in $M(\underline{X})$:

$$\frac{\Gamma \longrightarrow C \qquad \dfrac{\Delta \longrightarrow D \quad C, D \longrightarrow C\cdot D}{C, \Delta \longrightarrow C\cdot D}}{\Gamma, \Delta \longrightarrow C\cdot D} \; , \qquad \frac{\Lambda \longrightarrow B \qquad \dfrac{A/B, B \longrightarrow A \quad \Phi, A, \Psi \longrightarrow C}{\Phi, A/B, B, \Psi \longrightarrow C}}{\Phi, A/B, \Lambda, \Psi \longrightarrow C} \; .$$

PROPOSITION 2. Each proof in $M(\underline{X})$ is equivalent to the expansion of a proof in $G(\underline{X})$.

Proof. First we show, by induction on the number of symbols in A, that $A \longrightarrow A$ is a theorem in $G(\underline{X})$ whose proof, when expanded in $M(\underline{X})$, is equivalent to $1_A : A \longrightarrow A$.

Clearly $1_X: X \longrightarrow X$ is equivalent to $M_{1_X}: X \longrightarrow X$, since $\delta(\underline{X})$ is a functor.

$C \cdot D \longrightarrow C \cdot D$ may be proved thus in $G(\underline{X})$:

$$\frac{\dfrac{C \xrightarrow{P} C \quad D \xrightarrow{Q} D}{C,D \longrightarrow C \cdot D}}{C \cdot D \longrightarrow C \cdot D} \, ,$$

and we may assume, by inductional assumption, that $P \equiv 1_C$ and $Q \equiv 1_D$. Expanding in $M(\underline{X})$, we see that the above is equivalent to

$$\frac{\dfrac{C \xrightarrow{1_C} C \quad \dfrac{D \xrightarrow{1_D} D \quad C,D \longrightarrow C \cdot D}{C,D \longrightarrow C \cdot D}}{C,D \longrightarrow C \cdot D}}{C \cdot D \longrightarrow C \cdot D} \equiv \frac{C,D \longrightarrow C \cdot D}{C \cdot D \longrightarrow C \cdot D} \, ,$$

since $F(\underline{X})$ is a multicategory. Finally, the last proof is equivalent to $1_{C \cdot D}: C \cdot D \longrightarrow C \cdot D$, since $[C \cdot D;\ C \cdot D] \longrightarrow [C,D;\ C \cdot D]$ is one-one and onto.

$C/D \longrightarrow C/D$ may be proved thus in $G(\underline{X})$:

$$\frac{\dfrac{D \xrightarrow{Q} D \quad C \xrightarrow{P} C}{C/D,D \longrightarrow C}}{C/D \longrightarrow C/D} \, ,$$

and we may assume, by inductional assumption, that $P \equiv 1_C$ and $Q \equiv 1_D$.

Expanding in $M(\underline{X})$, we see that the above is equivalent to

$$\frac{D \xrightarrow{1_D} D \qquad \dfrac{C/D,D \longrightarrow C \qquad C \xrightarrow{1_C} C}{C/D,D \longrightarrow C}}{\dfrac{C/D,D \longrightarrow C}{C/D \longrightarrow C/D}} \quad \equiv \quad \frac{C/D,D \longrightarrow C}{C/D \longrightarrow C/D} \ ,$$

since $F(\underline{X})$ is a multicategory. Finally, the last proof is equivalent to $1_{C/D}: C/D \longrightarrow C/D$, since $[C/D;\ C/D] \longrightarrow [C/D,D;\ C]$ is one-one and onto.

$C\backslash D \longrightarrow C\backslash D$ is treated similarly, thus completing the induction proof that $1_A: A \longrightarrow A$ is equivalent to a proof in $G(\underline{X})$.

$A,B \longrightarrow A\cdot B$ may be proved thus in $G(\underline{X})$:

$$\frac{A \longrightarrow A \qquad B \longrightarrow B}{A,B \longrightarrow A\cdot B} \ .$$

Expanding in $M(\underline{X})$ and using the above result, we see that this is equivalent to

$$\frac{A \xrightarrow{1_A} A \qquad \dfrac{B \xrightarrow{1_B} B \qquad A,B \longrightarrow A\cdot B}{A,B \longrightarrow A\cdot B}}{A,B \longrightarrow A\cdot B} \quad \equiv \quad A,B \longrightarrow A\cdot B \ ,$$

since $F(\underline{X})$ is a multicategory.

$A/B,B \longrightarrow A$ and $B,B\backslash A \longrightarrow A$ are treated similarly. There remains the cut, which will now be dealt with. Let us, for the moment, assume that the cut has been added to the rules of $G(\underline{X})$.

CUT ELIMINATION THEOREM. Every proof in $G(\underline{X})$ with cut is equivalent to a proof without cut.

Proof. We assume that P and Q are proofs of $G(\underline{X})$, expanded into proofs of $M(\underline{X})$, and want to show that the cut

$$\frac{\Lambda \xrightarrow{Q} A \qquad \Phi,A,\Psi \xrightarrow{P} B}{\Phi,\Lambda,\Psi \longrightarrow B}$$

is equivalent to a proof with cuts of smaller degree. By the <u>degree</u> of the exhibited cut we mean the total number of occurrences of the symbols I, $\cdot$, $/$, and $\backslash$ in Φ, Λ, Ψ, A, B.

Proof. We distinguish four cases.

Case 1. $P = M_f$ and $Q = M_g$. Then

$$\frac{\Lambda \xrightarrow{M_g} X \qquad \Phi,X,\Psi \xrightarrow{M_f} Y}{\Phi,\Lambda,\Psi \longrightarrow B} \equiv \Phi,\Lambda,\Psi \xrightarrow{M_h} B ,$$

where h is the multimap in $\underline{X}$ obtained by substituting g into f, since $\delta(\underline{X})$ is a functor.

Case 2. The last step of Q does not introduce the main operation symbol of A. For example, assume that it is $I \longrightarrow$, then we claim that

$$\dfrac{\dfrac{\Gamma,\Delta \xrightarrow{R} A}{\Gamma,I,\Delta \longrightarrow A} \quad \Phi,A,\Psi \xrightarrow{P} B}{\Phi,\Gamma,I,\Delta,\Psi \longrightarrow B} \equiv \dfrac{\dfrac{\Gamma,\Delta \xrightarrow{R} A \quad \Phi,A,\Psi \xrightarrow{P} B}{\Phi,\Gamma,\Delta,\Psi \longrightarrow B}}{\Phi,\Gamma,I,\Delta,\Psi \longrightarrow B} ,$$

where the cut on the right has smaller degree. To verify this, observe that

$$\dfrac{\Gamma,\Delta \longrightarrow A}{\Gamma,I,\Delta \longrightarrow A} , \quad \dfrac{\emptyset \longrightarrow I \quad \Gamma,I,\Delta \longrightarrow A}{\Gamma,\Delta \longrightarrow A}$$

must by "inverses" in $M(\underline{X})$, in view of the isomorphism $[\Gamma,I,\Delta;A] \longrightarrow [\Gamma,\Delta;A]$. Hence, our task is equivalent to showing that

$$\dfrac{\emptyset \longrightarrow I \quad \dfrac{\Gamma,I,\Delta \longrightarrow A \quad \Phi,A,\Psi \longrightarrow B}{\Phi,\Gamma,I,\Delta,\Psi \longrightarrow B}}{\Phi,\Gamma,\Delta,\Psi \longrightarrow B} \equiv \dfrac{\dfrac{\emptyset \longrightarrow I \quad \Gamma,I,\Delta \longrightarrow A}{\Gamma,\Delta \longrightarrow A} \quad \Phi,A,\Psi \longrightarrow B}{\Phi,\Gamma,\Delta,\Psi \longrightarrow B} .$$

But this follows from the associative law for multicategories.

Case 3. The last step of P does not introduce the main operation symbol of A. For example, assume that it is $I \longrightarrow$, then we claim that

$$\dfrac{\Lambda \xrightarrow{Q} A \quad \dfrac{\Gamma,\Delta,A,\Psi \xrightarrow{R} B}{\Gamma,I,\Delta,A,\Psi \longrightarrow B}}{\Gamma,I,\Delta,\Lambda,\Psi \longrightarrow B} \equiv \dfrac{\dfrac{\Lambda \xrightarrow{Q} A \quad \Gamma,\Delta,A,\Psi \xrightarrow{R} B}{\Gamma,\Delta,\Lambda,\Psi \longrightarrow B}}{\Gamma,I,\Delta,\Lambda,\Psi \longrightarrow B} ,$$

where the cut on the right has smaller degree. This is treated similarly to Case 2 above, using the commutative law for multicategories.

Case 4. The last steps of both P and Q introduce the main operation symbols of A. For example, assume they are $\longrightarrow I$ and $I \longrightarrow$, then we claim that

$$\frac{\emptyset \longrightarrow I \qquad \dfrac{\Phi,\Psi \longrightarrow B}{\Phi,I,\Psi \longrightarrow B}}{\Phi,\Psi \longrightarrow B} \equiv \Phi,\Psi \longrightarrow B \ .$$

This is an immediate consequence of the isomorphism $[\Phi,I,\Psi;B] \longrightarrow [\Phi,\Psi;B]$.

The reader may object that the examples chosen under cases 2 to 3 are hardly typical. However, all other subcases have already been treated in Deductive Systems and Categories (I).

By a testing multicategory we shall mean a category with $n+1$ objects $1, \ldots, n+1$ and only one non-identity multimap $1, \ldots, n \longrightarrow n+1$. By the generality of a proof in $M(\underline{X})$ we mean the set of all pairs $(\underline{X}',H)$ where $\underline{X}'$ is a finite direct sum of testing multicategories and $H: \underline{X}' \longrightarrow \underline{X}$ is a functor such that there exists a proof P' in $M(\underline{X}')$ with $H(P') = P$.

PROPOSITION 3. Equivalent proofs in $M(\underline{X})$ have the same generality.

This may easily be shown by induction on the length of the proof that the given proofs are equivalent.

PROPOSITION 4. The functor $\delta(\underline{X}): \underline{X} \longrightarrow UF(\underline{X})$ defined by

$$\delta(\underline{X})(X) = X \ , \quad \delta(\underline{X})(F) = [M_f] \ ,$$

is full and faithful.

The proof is similar to that of the corresponding result for standard constructions and will be omitted.

COHERENCE THEOREM FOR BMM. Two proofs in $M(\underline{X})$ are equivalent only if they have the same generality.

Proof. In view of Proposition 2, it suffices to show that two proofs $P,Q: \Lambda \longrightarrow B$ in $G(\underline{X})$ with the same generality are equivalent. Look at all the axioms appearing in P from left to right. Suppose the i-th axiom involves the multimap

$$f_i: X_{i,1}, \ldots, X_{i,n(i)} \longrightarrow X_{i,n(i)+1} \ .$$

Let $\underline{X}'$ be the direct sum of the testing multicategories whose non-identity multimaps are

$$1, \ldots, n(i) \longrightarrow n(i) + 1 \ .$$

If K_i is the injection of the i-th testing multicategory into $\underline{X}'$, let $H: \underline{X}' \longrightarrow \underline{X}$ be determined by

$$HK_i(j) = X_{i,j}, \ HK_i(\longrightarrow) = f_i \ .$$

Now generalize P to a proof P' in $G(\underline{X}')$, replacing f_i by the i-th testing multimap. Then $H(P') = P$. Since P is assumed to have the same generality as Q, we have a proof Q^* in $M(\underline{X})$ such that $H(Q^*) = Q$. Now, by Proposition 2, $Q^* \equiv Q'$ for some proof Q' in $G(\underline{X})$. Therefore, also $H(Q') \equiv Q$, and $P \equiv Q$ will follow from $P' \equiv Q'$. That this is so will be shown in a lemma.

LEMMA. Let $\underline{X}$ be a finite direct sum of testing multicategories and assume that $\Lambda \longrightarrow B$ is a formula in $G(\underline{X})$ in which no object of $\underline{X}$ occurs more than once. Then any two proofs P and Q of $\Lambda \longrightarrow B$ in $G(\underline{X})$ are equivalent.

Proof. We roll up our sleeves and proceed by induction on the number of operation symbols in $\Lambda \longrightarrow B$. After disposing of the preliminary cases $P = M_f$, $Q = M_g$, and $P = \ \longrightarrow I$, $Q = \ \longrightarrow I$, we consider the following.

Case 1. Suppose the last step of P is $I \longrightarrow$ and the last step of Q is $\longrightarrow \cdot$. Then, say

$$P = \frac{\Gamma,\Theta,\Delta \xrightarrow{P'} C\cdot D}{\Gamma,I,\Theta,\Delta \longrightarrow C\cdot D}\ , \quad Q = \frac{\Gamma,I,\Theta \xrightarrow{Q'} C \qquad \Delta \xrightarrow{Q''} D}{\Gamma,I,\Theta,\Delta \longrightarrow C\cdot D}\ .$$

Now we have in $M(\underline{X})$

$$\frac{\emptyset \longrightarrow I \qquad \Gamma,I,\Theta \xrightarrow{Q'} C}{\Gamma,\Theta \longrightarrow C}\ ,$$

hence $\Gamma,\Theta \longrightarrow C$ also has a proof Q^* in $G(\underline{X})$.

By inductional assumption we may put

$$P' \equiv \dfrac{\Gamma,\Theta \xrightarrow{Q^*} C \qquad \Delta \xrightarrow{Q''} D}{\Gamma,\Theta,\Delta \longrightarrow C\cdot D}\ , \qquad Q' \equiv \dfrac{\Gamma,\Theta \xrightarrow{Q^*} C}{\Gamma,I,\Theta \longrightarrow C}\ .$$

Thus, we wish to show that

$$\dfrac{\dfrac{\Gamma,\Theta \longrightarrow C \qquad \Delta \longrightarrow D}{\Gamma,\Theta,\Delta \longrightarrow C\cdot D}}{\Gamma,I,\Theta,\Delta \longrightarrow C\cdot D} \equiv \dfrac{\dfrac{\Gamma,\Theta \longrightarrow C}{\Gamma,I,\Theta \longrightarrow C} \qquad \Delta \longrightarrow D}{\Gamma,I,\Theta,\Delta \longrightarrow C\cdot D}\ .$$

Expanding these proofs in $M(\underline{X})$, we claim that

$$\dfrac{\dfrac{\Gamma,\Theta \to C \qquad \dfrac{\Delta \to D \quad C,D \to C\cdot D}{C,\Delta \to C\cdot D}}{\Gamma,\Theta,\Delta \to C\cdot D}}{\Gamma,I,\Theta,\Delta \to C\cdot D} \equiv \dfrac{\dfrac{\Gamma,\Theta \to C}{\Gamma,I,\Theta \to C} \qquad \dfrac{\Delta \to D \quad C,D \to C\cdot D}{C,\Delta \to C\cdot D}}{\Gamma,I,\Theta,\Delta \to C\cdot D}\ .$$

In view of the isomorphisms $[\Gamma,I,\Theta;C] \longrightarrow [\Gamma,\Theta;C]$ and $[\Gamma,I,\Theta,\Delta;C] \longrightarrow [\Gamma,\Theta,\Delta;C]$ this is tantamount to showing that

$$\dfrac{\dfrac{\emptyset \to I \qquad \Gamma,I,\Theta \to C}{\Gamma,\Theta \to C} \qquad \dfrac{\cdot\ \cdot\ \cdot\ \cdot\ \cdot}{C,\Delta \to C\cdot D}}{\Gamma,\Theta,\Delta \to C\cdot D} \equiv \dfrac{\emptyset \to I \qquad \dfrac{\Gamma,I,\Theta \to C \qquad \dfrac{\cdot\ \cdot\ \cdot\ \cdot\ \cdot}{C,\Delta \to C\cdot D}}{\Gamma,I,\Theta,\Delta \to C\cdot D}}{\Gamma,\Theta,\Delta \to C\cdot D}\ ,$$

and this holds by associativity.

Case 2. Suppose the last step of P is $I \longrightarrow$ and the last step of Q is $\longrightarrow /$. Then say

$$P = \dfrac{\Gamma,\Delta \xrightarrow{P'} A/B}{\Gamma,I,\Delta \longrightarrow A/B}\ , \qquad Q = \dfrac{\Gamma,I,\Delta,B \xrightarrow{Q'} A}{\Gamma,I,\Delta \longrightarrow A/B}\ .$$

As before, we use the inductional assumption to show that both premises may equivalently be inferred from $\Gamma,\Delta,B \longrightarrow A$.

In view of the fundamental isomorphisms of a biclosed monoidal multicategory, one wants to verify that

$$\dfrac{\emptyset \to I \quad \dfrac{\Gamma,I,\Delta \to A/B \quad A/B,B \to A}{\Gamma,I,\Delta,B \to A}}{\Gamma,\Delta,B \to A} \equiv \dfrac{\dfrac{\emptyset \to I \quad \Gamma,I,\Delta \to A/B}{\Gamma,\Delta \to A/B} \quad A/B,B \to A}{\Gamma,\Delta,B \to A} ,$$

and this holds by associativity.

Cases 3 and 4. Suppose the last step of P is $I \longrightarrow$ and the last step of Q is $I \longrightarrow$ or $\cdot \longrightarrow$. These cases are treated similarly.

Case 5. Suppose the last step of P is $I \longrightarrow$ and the last step of Q is $/ \longrightarrow$. There are three subcases, only one of which we shall consider as an example.

$$P = \dfrac{\Phi,\Theta,A/B,\Lambda,\Psi \longrightarrow C}{\Phi,I,\Theta,A/B,\Lambda,\Psi \longrightarrow C} , \quad Q = \dfrac{\Lambda \longrightarrow B \quad \Phi,I,\Theta,A,\Psi \longrightarrow C}{\Phi,I,\Theta,A/B,\Lambda,\Psi \longrightarrow C} .$$

Using the inductional assumption as above, one wants to show that

$$\dfrac{\dfrac{\Lambda \longrightarrow B \quad \Phi,\Theta,A,\Psi \longrightarrow C}{\Phi,\Theta,A/B,\Lambda,\Psi \longrightarrow C}}{\Phi,I,\Theta,A/B,\Lambda,\Psi \longrightarrow C} \equiv \dfrac{\Lambda \longrightarrow B \quad \dfrac{\Phi,\Theta,A,\Psi \longrightarrow C}{\Phi,I,\Theta,A,\Psi \longrightarrow C}}{\Phi,I,\Theta,A/B,\Lambda,\Psi \longrightarrow C} .$$

We omit the easy verification.

There are essentially nine other cases. These have already been discussed in Deductive Systems and Categories (I). Unfortunately the five cases considered here are not at all typical of the difficulties that may arise. For example, the special assumptions about the multicategory $\underline{X}$ have not yet been used. To deal with the remaining cases, three lemmas were required. The reader who wishes to take the trouble to look into these matters should be warned that one of these lemmas (Lemma 1), must be modified in the present context as follows:

If $\Phi, \Lambda, \Psi \longrightarrow B$ is provable in $G(\underline{X})$ and Λ is prime to B, then $\Lambda \longrightarrow I$ and $\Phi, \Psi \longrightarrow B$ are theorems.

The definition of "prime" must also be modified: We say that Λ is prime to Δ in $G(\underline{X})$, provided $\underline{X}$ contains no multimap $f: X_1, \ldots, X_n \longrightarrow Y$ with some X_i occurring in a term of Λ and Y occurring in a term of Δ.

Just to give the reader an idea, let us discuss one further case. Suppose the last steps of P and Q are $\longrightarrow \cdot$, then we may put

$$P = \frac{\Gamma \xrightarrow{P'} C \quad \Theta, \Delta \xrightarrow{P''} D}{\Gamma, \Theta, \Delta \longrightarrow C \cdot D}, \quad Q = \frac{\Gamma, \Theta \xrightarrow{Q'} C \quad \Delta \xrightarrow{Q''} D}{\Gamma, \Theta, \Delta \longrightarrow C \cdot D}.$$

We assert that Θ is prime to D. For, otherwise, there is a multimap $f: X_1, \ldots, X_n \longrightarrow Y$ in $\underline{X}$ with X_i in Θ and Y in D. Now Y must be introduced into the proof of $\Delta \longrightarrow D$ through some axiom $Y_1, \ldots, Y_p \longrightarrow Z$ $(Y = Y_j)$ or $Z_1, \ldots, Z_q \longrightarrow Y$.

The first possibility is excluded by the definition of $\underline{X}$ as a direct sum: the components can have nothing in common. In the second case all Z_j occur in Δ, D. As we have assumed that no object of $\underline{X}$ occurs more than once, X_i is not among the Z_j, and, therefore, Y would be in two disjoint components of $\underline{X}$, again a contradiction.

Now we may invoke the above mentioned lemma to infer that both $\Theta \longrightarrow I$ and $\Delta \longrightarrow D$ are theorems in $G(\underline{X})$. By inductional assumption,

$$P'' \equiv \frac{\Theta \longrightarrow I \qquad \dfrac{\Delta \longrightarrow D}{I,\Delta \longrightarrow D}}{\Theta,\Delta \longrightarrow D} \quad , \quad Q' \equiv \frac{\Theta \longrightarrow I \qquad \dfrac{\Gamma \longrightarrow C}{\Gamma, I \longrightarrow C}}{\Gamma,\Theta \longrightarrow C}$$

Substituting these proofs into P and Q, we may show that both P and Q are equivalent to

$$\frac{\Theta \longrightarrow I \qquad \dfrac{\dfrac{\Gamma \longrightarrow C \qquad \dfrac{\Delta \longrightarrow D \quad C,D \longrightarrow C\cdot D}{C,\Delta \longrightarrow D}}{\Gamma,\Delta \longrightarrow C\cdot D}}{\Gamma, I,\Delta \longrightarrow C\cdot D}}{\Gamma,\Theta,\Delta \longrightarrow C\cdot D}$$

in $M(\underline{X})$. We omit the details.

One final word about the usefulness of the above results. Proposition 2 tells us how to find all proofs of $\Lambda \longrightarrow B$ in $M(\underline{X})$ up to equivalence. Proposition 3 and the coherence theorem tell us exactly when two proofs of $\Lambda \longrightarrow B$ are equivalent.

Since multimaps in $F(\underline{X})$ are equivalence classes of proofs in $M(\underline{X})$, we thus have a method for computing $[\Lambda;B]$ in $F(\underline{X})$. This assumes, of course, that the sets $[X_1, \ldots, X_n; Y]$ in $\underline{X}$ are known.

REFERENCES

For additional items see the references in 4.

[1] Beck, J., "The tripleableness theorem", manuscript.

[2] Benabou, J., "Algèbre élémentaire dans les catégories avec multiplication", *C. R. Acad. Sc. Paris*, 258; pp. 771-774, (1964).

[3] Benabou, J., "Catégories relatives", *C. R. Acad. Sc. Paris*, 260; pp. 3824-3827, (1965).

[4] Lambek, J., "Deductive systems and categories (I)", *Math. Systems Theory*, to appear.

[5] Paré, R., "Absolute coequalizers", in this volume.

POSSIBLE PROGRAMS FOR CATEGORISTS

by

Saunders MacLane

1. INTRODUCTION

Communication among Mathematicians is governed by a number of unspoken rules. One of these specifies that a Mathematician should talk about explicit theorems or concrete examples, and not about speculative programs. I propose to violate this excellent rule.

Category theory today is both a specialty and a generality. Specialities are the many particular fields in which current Mathematical knowledge and folklore develops; a new specialty arises in a field when the knowledge in that field and its prospects of further development demand full time workers. In the last six or eight years, category theory has become a flourishing specialty. On the other hand, a generality is an activity designed to encompass within suitable formal concepts a variety of results from special disciplines; it exists to help organize Mathematical knowledge. No one collection of formal concepts has for long sufficed as a generality. Indeed, historically there has been a succession of differing generalities: Mathematical Logic, Set Theory, General Topology, Abstract Algebra,

Lattices, Structures (in the sense of Bourbaki), and Categories. Today the proliferation of Mathematical ideas and Mathematical specialities makes it all the more imperative that good generalities be developed vigorously. The title of this conference speaks of Category Theory and its Applications, and the organization of this conference emphasizes the applications to Algebraic Topology, Homological Algebra, and Algebraic Geometry. But Category Theory, as a generality, has many other potential applications. Here we shall explore a few of them.

2. NEW GENERAL CONCEPTS

Our vital first question is: What other generalities should develop, above and beyond those of Categorical Algebra? I refer here not to refined types of categories or systems like categories, but to such brand new notions as may be suited to the codification of knowledge. I do not venture to predict what the new notions will be, but only that their development will require sharp attention to the multiplicity of Mathematical results needing comprehension.

3. THE REFINEMENT OF CONCEPTS

For their efficient use as generalities, the concepts of category theory require a long period of polishing, perfecting, and adaptation to put them in the most effective form for general use. For example, the notion of an abelian category required time, adjustment, and application to sheaf theory before it became really workable.

The basic notions of adjoint functors have required a much longer shakedown period. To begin with, the notion is much like that of adjoint linear transformation, but the successful abstract codification of the latter idea about 1929 by Stone and von Neumann did not directly lead to the idea of adjoint functor. The proximate sources of this idea are the universal constructions of Samuel and Bourbaki (1948) and the basic paper of D. M. Kan (1958) defining adjoint functors. Still, the interplay between these two aspects of adjointness (universality and natural isomorphism of hom sets) took another six or eight years to shake down into fully flexible form. The development of many other concepts may illustrate the same point, that the use of category theory as an effective generality requires a suppleness and a perfection (not an elaboration) in the utilization of concepts.

4. CATEGORIES WITH ADDED STRUCTURE

As an instrument of generality, a category will often come equipped with suitable added elements of structure. The present scene presents almost too many and varied such elements. Here are a few which seem likely to live, though some may survive better when they are more suitably renamed. Ehresmann's double categories are equipped with two sorts of composition ("vertical" and "horizontal"): They have a variety of realizations, including some recent applications by Paul Palmquist (not yet published) to the theory of adjunctions. The multiplicative categories (= monoidal categories) are categories equipped with an abstract tensor product operation, associative up to a coherent family of natural isomorphisms; they are omnipresent. While a category consists of objects and arrows (morphisms), with composition of arrows, Benabou's bicategories consist of objects, arrows, and two-cells, with compositions both of arrows and of two-cells, following Godement's well known five rules of functorial calculus. The composition of arrows in a bicategory is associative only up to coherent isomorphisms, so the multiplicative category is included as the special case of a bicategory with just one object. The further development of this idea is most suggestive, but sufficiently complex, so that it may be a while before it is clear whether the straight-forward generalization to "tricategories" is viable.

A fundamental role is played by those multiplicative categories $\mathbb{M}$ where for each object the functor $X \longmapsto A \otimes X$ has a right adjoint (an internal hom functor $Y \longmapsto \mathrm{Hom}(A,Y)$). Modifying an earlier terminology of Linton's, such categories might best be called autonomous categories. Note, in particular, that their definition requires no reference to an underlying category of sets. A cartesian closed category $\mathbb{C}$ is a special case; it is a category with finite products in which each $X \longmapsto A >< X$ has a right adjoint; the lectures of Lawvere at this conference indicate the importance of this notion (the terminology, due to Eilenberg-Kelly, is not fortunate, since their so-called closed categories do not exist in nature). With autonomous categories one naturally comes to the biautonomous categories--multiplicative categories where the abstract tensor product is not (known to be) commutative, but where both $X \longmapsto A \otimes X$ and $X \longmapsto X \otimes A$ have right adjoints. One can expect that these notions still require a shake-down.

5. AXIOMATIC MANIFOLDS

Category theory originated in certain questions of algebraic topology, and hitherto its applications have tended to cluster there and in algebraic geometry. In the future, this may not be the case; there might develop extensive applications in other fields of Mathematics. I speculate that this may be the case in differential geometry, a subject which has not yet undergone the rationalization experienced by topology and algebraic geometry. For instance, one might search for axioms on the category of all C^∞ - manifolds or on this category enlarged to contain suitable infinitessimal objects (as in Lawvere's unpublished work on categorical dynamics).

6. UNIFOLDS

It is an elementary observation that categorical methods have had no effect upon analysis, classical or abstract. There may yet be some notions of analysis open to such applications. One possible place is the treatment of those local properties of functions $f(x_1,\dots,x_m)$ of several variables $x_1,\dots,x_m$ which depend not on the explicit choice of the variables, but on the local geometrical properties of a neighborhood U with coordinates $x_1,\dots,x_m$. These are, if you will, properties of the smooth manifold in which U is a neighborhood. However, there are many such properties which are strictly local and so need no global reference to a manifold. The appropriate notion may be that of a "unifold". A C^m- unifold (a unifold of class m) is defined to be a set U together with a set F of real-valued functions $f: U \longrightarrow \mathbb{R}$, for which there exists a positive integer n, a list $x_1,\dots,x_n$ of functions in F and an open set U' in $\mathbb{R}^n$ all such that $u \longmapsto (x_1(u),\dots,x_n(u))$ is a bijection of U to U' for which the functions f in F are precisely the composites $g(x_1,\dots,x_n)$ with g a C^m function on U' to $\mathbb{R}$. In other words, a unifold is just an open set in $\mathbb{R}^n$, but without the choice of any one coordinate system. Those local properties of functions of several variables which are independent of the choice of variables can be naturally stated as properties of unifolds. For instance, this is the case for the theory of characteristics of first order partial differential equations.

7. AXIOMS ON THE CATEGORICAL LEVEL

The axiomatic method is usually taken to be a way of studying a group or a topological space by axioms valid in any one group or in any one topological space. An alternative now open is that of considering as axioms those properties holding for the category of <u>all</u> groups or of <u>all</u> topological spaces. This prospect was first envisaged by Lawvere in his axioms for the category of sets (<u>An Elementary Theory of the Category of Sets</u>) and for the category of all categories (<u>The Category of Categories as a Foundation for Mathematics</u>). Some additional such "global" axiom systems have been studied by Schlomiuk and others. It would seem that there might be many such systems--not just the one, noted above, of axioms on the category of all smooth manifolds.

8. UNIVERSAL ALGEBRA

The attractive possible use of category theory in universal algebra has not yet been realized. Let us recall the basic idea, as formulated in Lawvere's thesis. An algebraic system is usually described as a set together with certain unary, binary, ternary,... operations satisfying specified identities. For example, a group is a set G together with a binary operation (multiplication), a unary operation (the inverse), and a nullary operation (the group identity) which together satisfy the well-known group axioms. However, these three operations are clearly not the only ones at hand. For each group-word $w(x_1,\ldots,x_n)$ in n letters x_i (in the corresponding free group) there is the n-ary operation $g_1,\ldots,g_n \longmapsto w(g_1,\ldots,g_n)$ on G. The theory of groups really deals simultaneously with all these operations. By speaking of m-tuples of n-ary operations as morphisms $n \longrightarrow m$, the theory of groups becomes a category (with objects the non-negative integers). This is an invariant description of the theory of groups. It compares to the classical description of group theory as the study of three particular operations (multiplication, inverse, and identity) just as the notion of an individual abstract group compares to a presentation of that group by particular generators and relations. The important task of rewriting the results of universal algebra in this appropriate invariant language still remains to be done.

9. FOUNDATIONS

It is an open scandal that the classical method of applying Zermelo-Fraenkel set theory as a foundation for all practise of Mathematics is no longer adequate to the practise of Category Theory. The device of having both large and small categories in some Gödel-Bernays set theory was a convenient arrangement when it was first proposed by Eilenberg-MacLane 23 years ago, but it no longer convenes for functor categories (with large domain category) or for the category of all categories as used in the theory of fibered categories or in Benabou's profunctors. The alternative arrangement of categories in a Grothendieck Universe has been effective for getting on with the development of Mathematics, but it introduces assumptions as to inaccessible cardinals which palpably have nothing to do with the case, and it leaves unsettled (as yet) a variety of questions of the possible effects resulting from a shift of universe.

What should we conclude? The happy security provided by one "monolithic" foundation has been lost. First Principia Mathematica and then Zermelo-Fraenkel had this monolithic character, that all working Mathematics could be formulated within one system. This provided a convenient division of labor, between Mathematicians who just "used" the system (usually in a "naive" form) and the Logician who investigated within the system various classical problems. This paradise is irretrievably lost; it is high time that open-minded young Mathematicians set to work to construct a new one--perhaps less monolithic.

10. CONCLUSION

This brief survey does not pretend to cover all the profitable current problems in Category Theory and its use as a generality. A much better coverage is provided by the various specific talks given at this conference--and, we hope, by the other new and different ideas which lie in the future.

ABSOLUTE COEQUALIZERS

by

Robert Paré

The concept of split coequalizer plays an important role in Beck's tripleableness theorem (see [1]). Although this notion does have some justification in the fact that it is part of a contraction of a simplicial object, it is somewhat artificial.

In this paper we consider an equivalent formulation of Beck's theorem in which the notion of split coequalizer is replaced by the more natural notion of absolute coequalizer. Although the new conditions would appear, at first glance, to be more difficult to verify, it seems that all we need in practice is the absoluteness property. In many cases, the fact that we have no equation, but only absoluteness leads to more conceptual proofs.

In the first part of this paper, we give the necessary definitions and outline the proof of Beck's theorem from this point of view. Then a proof that the category of semigroups is tripleable over sets is sketched to show how these new conditions can be used.

The second part contains several characterizations of absolute coequalizers. Then it is shown that there exist absolute coequalizers which are not split.

I would like to thank my director of research, Professor Lambek, for his guidance and encouragement.

I. Definition 1: Let $\underline{A}$ be a category and let

$$Y_1 \underset{d_1}{\overset{d_o}{\rightrightarrows}} Y_o \xrightarrow{d} Y$$

be a coequalizer diagram in $\underline{A}$. $(d_o, d_1; d)$ is called an absolute coequalizer if for every category $\underline{C}$ and every functor $G : \underline{A} \longrightarrow \underline{C}$, $(G(d_o), G(d_1); G(d))$ is also a coequalizer.

Let $U : \underline{B} \longrightarrow \underline{A}$ be a functor.

Definition 2. Let $Y_1 \underset{d_1}{\overset{d_o}{\rightrightarrows}} Y_o$ be maps in $\underline{B}$.

(d_o, d_1) is U-absolute if there exist Z and $U(Y_o) \xrightarrow{d} Z$ in $\underline{A}$ such that $U(Y_1) \underset{U(d_1)}{\overset{U(d_o)}{\rightrightarrows}} U(Y_o) \xrightarrow{d} Z$ is an absolute coequalizer.

Definition 3: $\underline{B}$ has U-absolute coequalizers if each U-absolute pair of maps in $\underline{B}$ has a coequalizer (not necessarily absolute) in $\underline{B}$.

Definition 4: U preserves U-absolute coequalizers if whenever $Y_1 \rightrightarrows Y_\circ$ is U-absolute and has a coequalizer $Y_\circ \longrightarrow Y$ in $\underline{B}$, the canonical map $Z \longrightarrow U(Y)$ is an isomorphism.

Definition 5: U reflects U-absolute coequalizers if $Y_1 \rightrightarrows Y_\circ \longrightarrow Y$ being mapped into an absolute coequalizer by U implies $Y_1 \rightrightarrows Y_\circ \longrightarrow Y$ is a coequalizer (not necessarily absolute) in $\underline{B}$.

From now on let $U : \underline{B} \longrightarrow \underline{A}$ have a left adjoint $F : \underline{A} \longrightarrow \underline{B}$. This adjoint pair gives rise (see [6] or [2]) to a triple $\mathbf{T} = (T,\eta,\mu)$ on $\underline{A}$. Let $\underline{A}^{\mathbf{T}}$ be the category of $\mathbf{T}$-algebras and $U^{\mathbf{T}} : \underline{A}^{\mathbf{T}} \longrightarrow \underline{A}$ its underlying functor. There is a canonical map $\Phi : \underline{B} \longrightarrow \underline{A}^{\mathbf{T}}$ making the following diagram commute:

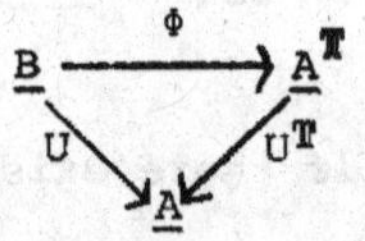

Definition 6: U is said to be tripleable if Φ has a left adjoint $\overline{\Phi}$ such that the adjunctions $1_{\underline{A}^{\mathbf{T}}} \longrightarrow \Phi\overline{\Phi}$ and $\overline{\Phi}\Phi \longrightarrow 1_{\underline{B}}$ are natural isomorphisms.

We can now state Beck's tripleability theorem.

Theorem: U is tripleable $\Longleftrightarrow$ U has a left adjoint, $\underline{B}$ has U-absolute coequalizers, and U preserves and reflects U-absolute coequalizers.

Remark: $Y_1 \overset{d_\circ}{\underset{d_1}{\rightrightarrows}} Y_\circ \xrightarrow{d} Y$ is a split coequalizer if there exist $Y \xrightarrow{h} Y_\circ$ and $Y_\circ \xrightarrow{h_1} Y_1$ such that the following equations are satisfied:

$$\begin{aligned} dd_\circ &= dd_1 \\ dh &= 1_Y \\ hd &= d_\circ h_1 \qquad (1) \\ d_1 h_1 &= 1_{Y_\circ} . \end{aligned}$$

Equations (1) imply that $(d_\circ, d_1; d)$ is a coequalizer, thus we conclude that split coequalizers are absolute.

If we replace the word absolute by the word split in the preceding definitions and theorem, we obtain the original form of the theorem.

Proof of theorem: Since split coequalizers are a special case of absolute ones, the sufficiency of the condition offers no difficulty.

Now, suppose that U is tripleable. We can assume that $\underline{B} = \underline{A}^{\mathbf{T}}$ and prove that $\underline{A}^{\mathbf{T}}$ has, and $U^{\mathbf{T}}$ preserves and reflects, $U^{\mathbf{T}}$-absolute coequalizers.

Let $(Y_1, \theta_1) \overset{d_\circ}{\underset{d_1}{\rightrightarrows}} (Y_\circ, \theta_\circ)$ be a $U^{\mathbf{T}}$-absolute pair of maps in $\underline{A}^{\mathbf{T}}$. We have the following absolute coequalizer in $\underline{A}$:

$$Y_1 \overset{d_\circ}{\underset{d_1}{\rightrightarrows}} Y_\circ \xrightarrow{d} Y .$$

Consider the following diagram:

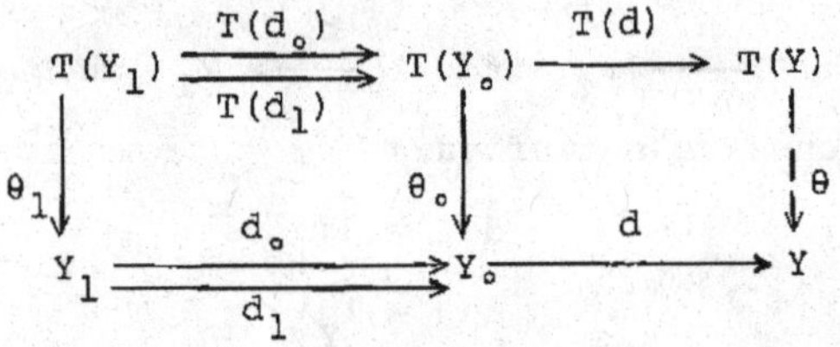

The upper and lower squares on the left commute since $d_\circ$ and d_1 are $\mathbb{T}$-homomorphisms. Thus,

$d\theta_\circ T(d_\circ) = dd_\circ\theta_1 = dd_1\theta_1 = d\theta_\circ T(d_1)$, but $(T(d_\circ), T(d_1); T(d))$ is a coequalizer; therefore there exists a unique $\theta : T(Y) \longrightarrow Y$ making the square on the right commute.

We shall prove that (Y,θ) is a $\mathbb{T}$-algebra. Consider the following cube:

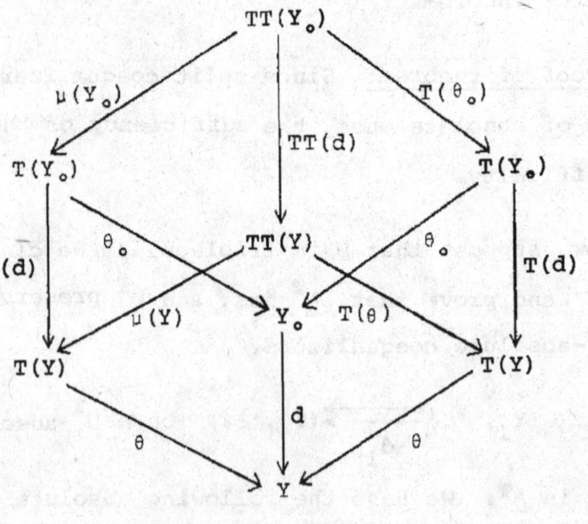

The top face commutes because of the $\theta_\circ$-associativity axiom. One lateral face commutes by naturality of μ, the other three by definition of θ. Since $(TT(d_\circ), TT(d_1); TT(d))$ is a coequalizer $TT(d)$ is epi, and thus the bottom face also commutes (see [7], p. 43). This proves the θ-associativity axiom.

To prove the θ-unitary axiom, consider:

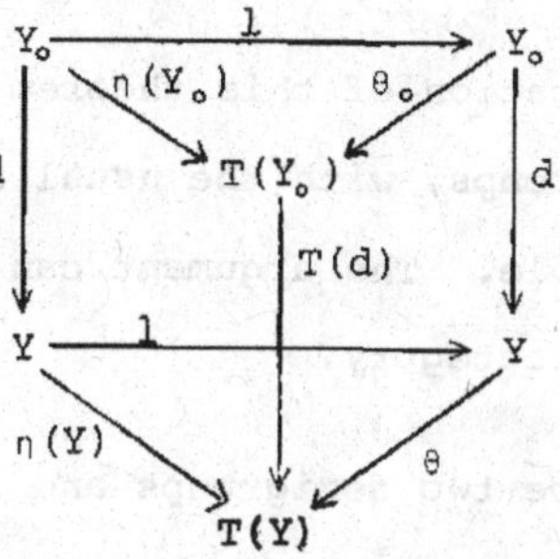

The top face commutes by the $\theta_\circ$-unitary axiom, one lateral face commutes by naturality of η, one by definition of θ, and the other trivially. Since d is epi the bottom face must also commute. (This diagram is actually a degenerate cube.)

We have proved that (Y,θ) is a $\mathbb{T}$-algebra, and by definition of θ, d is a $\mathbb{T}$-homomorphism.

Finally, $d = \text{coeq}\ (d_\circ, d_1)$ in $\underline{A}^{\mathbb{T}}$ for if a $\mathbb{T}$-homomorphism, x, coequalizes $(d_\circ, d_1)$ in $\underline{A}^{\mathbb{T}}$, it also does in $\underline{A}$, and thus there exists a unique z in $\underline{A}$ such that the following diagram commutes:

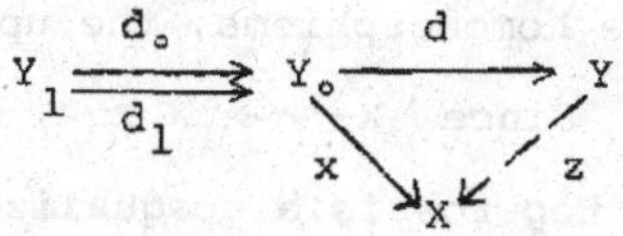

It is easily verified that z is a $\mathbb{T}$-homomorphism.

We have shown that $\underline{A}^{\mathbb{T}}$ has $U^{\mathbb{T}}$-absolute coequalizers. This construction also shows that $U^{\mathbb{T}}$ preserves and reflects $U^{\mathbb{T}}$-absolute coequalizers.

Q.E.D.

As an application of this theorem, we will show that the category of semigroups, with the usual underlying functor into sets, is tripleable. The argument can easily be carried out for any algebraic category.

Let Y_1, $Y_\circ$ be two semigroups and let $Y_1 \overset{d_\circ}{\underset{d_1}{\rightrightarrows}} Y_\circ$ be two homomorphisms. Assume that $Y_1 \overset{d_\circ}{\underset{d_1}{\rightrightarrows}} Y_\circ \xrightarrow{d} Y$ is an absolute coequalizer in sets. We must define a semigroup structure on Y, so that d is a homomorphism.

If m_1 and $m_\circ$ are the multiplications of Y_1 and $Y_\circ$ respectively, we have:

$$\begin{array}{ccccc} Y_1 \times Y_1 & \overset{(d_\circ,d_\circ)}{\underset{(d_1,d_1)}{\rightrightarrows}} & Y_\circ \times Y_\circ & \xrightarrow{(d,d)} & Y \times Y \\ \downarrow{\scriptstyle m_1} & & \downarrow{\scriptstyle m_\circ} & & \downarrow{\scriptstyle m} \\ Y_1 & \overset{d_\circ}{\underset{d_1}{\rightrightarrows}} & Y_\circ & \xrightarrow{d} & Y \end{array}$$

Because $d_\circ$ and d_1 are homomorphisms, the upper and lower squares on the left commute. Since $X \rightsquigarrow X \times X$, $f \rightsquigarrow (f,f)$ defines a functor, the top row is a coequalizer diagram. $dm_\circ(d_\circ, d_\circ) = dd_\circ m_1 = dd_1 m_1 = dm_\circ(d_1, d_1)$ thus there exists a unique $m : Y \times Y \longrightarrow Y$ making the square on the right commute.

To see that m is associative, consider the following cube:

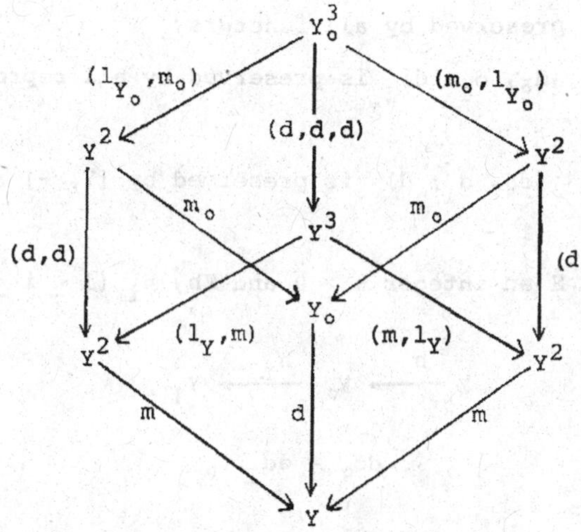

The top face commutes because m_o is associative, the lateral faces commute by definition of d, and (d,d,d) is epi, therefore the bottom must also commute. This shows that (Y,m) is a semigroup. By definition of m, d is a homomorphism.

It is easily verified that $d = \text{coeq}\,(d_o, d_1)$ in the category of semigroups. It is also easy to verify that the underlying functor preserves and reflects the right things. We conclude that the category of semigroups is tripleable over sets.

II. We shall now see several characterizations of absolute coequalizers.

Theorem: Let $Y_1 \overset{d_o}{\underset{d_1}{\rightrightarrows}} Y_o \xrightarrow{d} Y$ be a coequalizer in $\underline{A}$. The following statements are equivalent:

(i) $(d_o, d_1; d)$ is an absolute coequalizer, i.e. $(d_o, d_1; d)$ is preserved by all functors

(ii) $(d_o, d_1; d)$ is preserved by all representable functors

(iii) $(d_o, d_1; d)$ is preserved by $[Y, -]$ and $[Y_o, -]$

(iv) $\exists$ an integer $n \geq 0$ and $\exists h, h_i$ $(1 \leq i \leq n)$ maps in $\underline{A}$ such that

$$Y \xrightarrow{h} Y_o \xrightarrow{h_i} Y_1$$

verifying

$$dd_o = dd_1$$

$$dh = 1_Y$$

$$hd = d_{\nu(1)}h_1$$

$$d_{\nu(2)}h_1 = d_{\nu(3)}h_2$$

$$d_{\nu(4)}h_2 = d_{\nu(5)}h_3$$

$$\vdots$$

$$d_{\nu(2n)}h_n = 1_{Y_o}$$

where $\nu(i) = 0$ or 1.

Proof: (i) $\Longrightarrow$ (ii) $\Longrightarrow$ (iii) obvious.

(iii) $\Longrightarrow$ (iv) : $[Y, Y_1] \overset{[Y,d_o]}{\underset{[Y,d_1]}{\rightrightarrows}} [Y,Y_o] \xrightarrow{[Y,d]} [Y,Y]$

is a coequalizer diagram in sets. $[Y,d]$ is therefore onto. Let $h \in [Y,Y_\circ]$ such that $[Y,d](h) = 1_Y$, i.e. $dh = 1_Y$.

$$\text{Now} \quad [Y_\circ, Y_1] \underset{[Y_\circ,\, d_1]}{\overset{[Y_\circ,\, d_\circ]}{\rightrightarrows}} [Y_\circ, Y_\circ] \xrightarrow{[Y_\circ,\, d]} [Y_\circ, Y]$$

is also a coequalizer in sets.

$$[Y_\circ, d](hd) = dhd = d = [Y_\circ, d](1_{Y_\circ}),$$

thus, by the construction of the coequalizer in sets, $hd \sim 1_{Y_\circ}$ where $\sim$ is the equivalence relation generated by the relation

$$R = \{(d_\circ x, d_1 x) \mid x \in [Y_\circ, Y_1]\} .$$

Thus there exists an integer $n \geq 0$ and $\exists h_i : Y_\circ \longrightarrow Y_1$ $(i = 1, 2, 3, \ldots, n)$ such that

$$hd = d_{\nu(1)} h_1$$

$$d_{\nu(2)} h_1 = d_{\nu(3)} h_2$$

$$d_{\nu(4)} h_2 = d_{\nu(5)} h_3$$

$$\vdots$$

$$d_{\nu(2n)} h_n = 1_{Y_\circ}$$

where $\nu(i) = 0$ or 1. We obviously also have the relation $dd_\circ = dd_1$.

(iv) $\Longrightarrow$ (i) : The equations given in (iv) imply that $(d_\circ, d_1; d)$ is a coequalizer, and they are obviously preserved by any functor.

Q.E.D.

Remark 1: It can be shown that $(d_o, d_1; d)$ is an absolute coequalizer $\Longleftrightarrow$ $(d_o, d_1; d)$ is preserved by all full embeddings.

Remark 2: It is not necessary to assume that $(d_o, d_1; d)$ is a coequalizer but only that it is mapped into coequalizers by the functors $[Y, -]$, $[Y_o, -]$, and $[Y_1, -]$.

It is natural to ask if absolute coequalizers are different from split coequalizers. The answer is yes. Consider the diagram:

$$Y_1 \underset{d_o}{\overset{d_o}{\rightrightarrows}} Y_o \xrightarrow{1_{Y_o}} Y_o$$

$(d_o, d_o; 1_{Y_o})$ is an absolute coequalizer, but it is split $\Longleftrightarrow$ d_o is a split epi.

Definition: We shall say that an absolute coequalizer is n-split if there exist n maps d_i such that condition (iv) of the preceding theorem is verified.

Above is an example of a 0-split coequalizer. Split coequalizers are 1-split. It is readily seen that for $n > 0$, n-split implies $(n + 1)$-split.

We can show that for every integer $n \geq 0$, there exist n-split coequalizers which are not i-split for any $0 \leq i < n$.

Indeed, define $Y_o \underset{d_1}{\overset{d_o}{\rightrightarrows}} Y_1 \xrightarrow{d} Y$ in sets as follows:

$Y_o = Y_1 = \{1,2,3,\ldots, n+1\}$, $Y = \{1\}$

$d_o = 1_{Y_o}$, $d_1(i) = \max(i-1, 1)$, $d(i) = 1$. Now define $h : Y \longrightarrow Y_1$ by $h(1) = 1$ and define $h_k : Y_1 \longrightarrow Y_o$ by $h_k(i) = \max(i+k-n, 1)$. Then $h_{n-1}(i) = \max(i-1, 1)$ and we have the following results:

$$h_k = (h_{n-1})^{n-k}$$

$$d_1 = h_{n-1} .$$

It follows immediately that we have the relations:

$$dd_o = dd_1$$

$$dh = 1_Y$$

$$hd = d_1h_1 = (h_{n-1})^n$$

$$d_oh_1 = d_1h_2 = (h_{n-1})^{n-1}$$

$$\vdots$$

$$d_oh_n = 1_{Y_o} = (h_{n-1})^0 .$$

Thus $(d_o, d_1; d)$ is n-split.

On the other hand, if there exist h', h_i', $(i = 1, 2, \ldots, m)$ such that

$$h'd = d_{\nu(1)}h'_1$$

$$d_{\nu(2)}h'_1 = d_{\nu(3)}h'_2$$

$$\vdots$$

$$d_{\nu(2m)}h'_m = 1_{Y_o}$$

we can establish the following relations on the cardinalities of the images of the maps under consideration:

$$|\mathrm{Im}(d_{\nu(i)}h'_j)| = \begin{cases} |\mathrm{Im}(h'_j)| \\ \text{or} \\ |\mathrm{Im}(h'_j)| - 1 . \end{cases}$$

Thus in going from one equation to the following, the cardinality of the image can be changed by at most 1. But $|\mathrm{Im}(dh')| = 1$ and $|\mathrm{Im}(1_{Y_o})| = n + 1$, thus there must be at least $n + 1$ equations, i.e. $m \geq n$. We conclude that $(d_o, d_1; d)$ is not i-split for any $i < n$.

I am indebted to Michael Barr for suggesting the preceding example to replace a more complicated one.

REFERENCES

[1] Beck, J. "The Tripleableness Theorem" (Manuscript).

[2] Beck, J. "Triples, Algebras, and Cohomology", (*Dissertation*, (1967), Columbia University).

[3] Linton, F. E. J. "An Outline of Functorial Semantics", (Lecture Notes in Math, Springer - to appear).

[4] Linton, F. E. J. "Coequalizers in Categories of Algebras", (Lecture Notes in Math, Springer - to appear).

[5] Linton, F. E. J. "Applied Functorial Semantics I", (*Annali di Matematica*, to appear).

[6] Manes, E. G. "A Triple Miscellany: some aspects of the theory of algebras over a triple", (*Dissertation*, Wesleyan University, Middletown, Conn., 1967).

[7] Mitchell, B. *Theory of Categories*, Academic Press, New York, (1965).

GALOIS THEORY

by

Stephen S. Shatz*

INTRODUCTION

This will be a summary and discussion - mostly without proofs, or with barest indications of same - of my recent work on Galois Theory for field extensions. I was motivated by certain "obvious" applications of any such theory; and these applications suggest the boundary lines for the sort of theorems one needs.

Specifically, one wants to construct a "fundamental group scheme" for an arbitrary connected scheme, so as to classify the flat coverings instead of just classifying the *etale* coverings. For this one will need a Galois Theory for flat, finite coverings of schemes. Any Galois Theory will yield a description of the Brauer Group (at least in the affine case) if it is the "correct" theory; one wishes to study those elements split by a given flat covering. The third application is to classify <u>all</u> isogenies of a given group scheme. A blueprint for doing this is given by the classical

*The author wishes to acknowledge the support of N.S.F. G.P. 7654, Battelle Memorial Institute, and Stanford University during preparation of this manuscript.

separable case, and the height one case as done by Serre [3]. A good Galois Theory ought to put these together consistently, as well as enable one to handle the case of height $n > 1$.

To keep the exposition short, we shall deal only with the case of a finite, purely inseparable field extension K/k. Actually, generalization to the case of an arbitrary finite extension is easy, but requires more space. We want to emphasize that theorems stated below have been proven only in the field theoretic case - more work will be necessary to treat the general affine case.

1. THE GROUP SCHEME AUT, RING SCHEMES

Our point of view is that any Galois Theory should be a relationship between objects and "groups" of their automorphisms. Moreover, it should be functorial, and follow as closely as possible the model of the classical case.

It has been recognized for some time that, in many situations for which ordinary groups are inadequate and in which groups ought to play a part, the correct notion is a group scheme. So, quite naturally, we are lead to consider Grothendiecks automorphism functor, Aut, [1]. Recall that if X, Y are schemes, $\pi: Y \longrightarrow X$ a morphism, then Aut (Y/X) is that functor which associates to each X-scheme $\mathfrak{Z}$, the group of $\mathfrak{Z}$-automorphisms of $Y \times_X \mathfrak{Z}$. That is,

$$\underline{\mathrm{Aut}}\ (Y/X)\ (\mathfrak{Z}) = \mathrm{Aut}_{\mathfrak{Z}}\ (Y \times_X \mathfrak{Z}).$$

If Y and X are affine, say with rings Ω and Λ respectively, and if π makes Ω a finite, projective Λ-module of finite presentation, then one trivially sees that $\underline{\mathrm{Aut}}\ (Y/X) = \underline{\mathrm{Aut}}\ (\Omega/\Lambda)$ is representable by a group scheme of finite type over Λ. In particular, if K/k is an arbitrary finite field extension, then $\underline{\mathrm{Aut}}\ (K/k)$ is representable by an algebraic k-group scheme.

Observe, however, that $\underline{\mathrm{Aut}}\ (K/k)$ is a "large" scheme. Indeed, in the simplest non-trivial case: $K = k(\theta)$, $\theta^2 = t$, $t \notin k^2$, ch (k) = 2, one finds an exact sequence

$$0 \longrightarrow \alpha_2 \longrightarrow \underline{\mathrm{Aut}}\ (K/k) \longrightarrow \mathbb{G}_m \longrightarrow 0.$$

Now $\underline{\mathrm{Aut}}\ (K/k)$ is a functorial construction involving k, K, and should play some role in the Galois Theory. But we must turn k, K into functors as well, in view of the opening remarks of this section. For this, we can consider K as defining a "k-algebra scheme" (and k, the same), say $\underline{K}$, *via* the equation

$$\underline{K}\ (\mathfrak{Z}) = \text{Affine ring of } (\mathrm{Spec}\ K \times_k \mathfrak{Z})$$
$$= K \otimes_k (\text{Affine ring of } \mathfrak{Z}).$$

(For $\underline{k}$ one should read: $\underline{k}\ (\mathfrak{Z})$ = Affine ring of $\mathfrak{Z}$). Here $\mathfrak{Z}$ is any k-scheme; and one finds, trivially, that $\underline{K}$ and $\underline{k}$ are representable by ring schemes - more exactly, k-algebra schemes.

A simple verification shows that the natural injection $\underline{\text{Aut}}\ (K/k) \longrightarrow \underline{\text{Aut}}\ (\underline{K}/\underline{k})$ is an isomorphism - so the Aut does not have to be changed.

Now the supposed structure is clear: There should be a 1-1 lattice inverting correspondence between the subringschemes of $\underline{K}/\underline{k}$ and the (closed) subgroupschemes of $\underline{\text{Aut}}\ (K/k)$. But, this is not what our methods yield! We can only capture those subringschemes arising from subfields of the layer K/k. That there are others, the reader may easily check. For example, in the case described above: $K = k(\theta)$, $\theta^2 = t$, etc., for any ring A, $\underline{K}\ (A)$ is given by $K \otimes A = A \oplus A$ (as modules). If we look at all pairs $< x, y > \in \underline{K}\ (A)$ such that $y^2 = 0$, we obtain a ring scheme between $\underline{k}$ and $\underline{K}$, not arising from a subfield of the layer K/k (there are no non-trivial such!).

One more observation: The groupscheme $\underline{\text{Aut}}\ (K/k)$ is connected if and only if K/k is a purely inseparable extension.

2. THE GALOIS CORRESPONDENCE

Given a closed subgroupscheme $\underline{H}$ of $\underline{\text{Aut}}\ (K/k)$, we want to associate with it a subringscheme of $\underline{K}/\underline{k}$. The natural first attempt at a functor

$$A \rightsquigarrow (K \otimes A)^{\underline{H}(A)}$$

is not a functor, and has the wrong values. One must "localize" it or universalize it on the model of algebraic geometry

[2], so as to define

$$\underline{\text{Fix } H}\ (A) = \{\ \xi \in K \otimes A \mid \text{for every A-alg. B, the image of } \xi \text{ in } K \otimes B \text{ is left fixed by all of } \underline{H}\ (B)\}.$$

This definition $A \rightsquigarrow \text{Fix } \underline{H}\ (A)$ clearly yields a functor, Fix $\underline{H}$; moreover the process is quite obviously the exact analog of the classical process (suitably changed for functorial purposes).

Proposition. _For any extension_ K/k, _the functor_ Fix $\underline{H}$ _is a sheaf in the faithfully flat, quasi-compact topology over_ k; _in fact_, Fix $\underline{H}$ _is representable by the ring scheme corresponding to a unique subfield_ L _of the layer_ K/k. _This field_ L _is determined from the equation_

$$L = \text{Fix } \underline{H}\ (k),$$

and for each A, _one has_

$$\text{Fix } \underline{H}\ (A) = L \otimes A.$$

Observe the remarkable fact that the k-rational points of the scheme Fix $\underline{H}$ determine that scheme.

Now it may happen that distinct subgroupschemes $\underline{H}$, $\underline{H}'$ have the same Fix functors, i.e., their associated fields (= k-rational points of the Fix functor) are the same. For example, in the often cited case $k(\theta) = K$, $\theta^2 = t$, etc., one finds that $\underline{\text{Aut}}\ (K/k)$, μ_2, α_2, all have k as associated field. _Therefore, we decree that two subgroupschemes_ $\underline{H}$, $\underline{H}'$

will be called Galois equivalent if and only if Fix $\underline{H}$ (k) = Fix $\underline{H}'$ (k). We will write $\underline{H} \sim \underline{H}'$ in this case. The functor Fix then establishes a mapping from the Galois equivalence classes of subgroupschemes of Aut to the subfields of K/k.

Given a (closed) subringscheme $\underline{S}$ of $\underline{K/k}$, we want to associate with it a subgroupscheme of Aut (K/k). In view of our experience with Fix, and the classical definition, the following definition is completely natural:

$$\text{Inv. } \underline{S}\ (A) = \{\sigma \in \text{Aut}_A\ (K \otimes A) \mid \text{for every A-alg. B, the image of } \sigma \text{ in Aut}_B\ (K \otimes B) \text{ when restricted to } \underline{S}\ (B) \text{ shall be the identity.}\}.$$

Since $\text{Aut}_A\ (K \otimes A) = \underline{\text{Aut}}\ (K/k)\ (A)$, the functor Inv. $\underline{S}$ (it is clearly such) is a subfunctor of Aut (K/k).

Proposition. The functor Inv. $\underline{S}$ is representable by a closed subgroupscheme of Aut (K/k). If L is the smallest subfield of K/k such that $\underline{S} \subset \underline{L}$, then Inv. $\underline{S}$ = Inv. $\underline{L}$.

Clearly, both Inv and Fix invert order, and what remains is to show that they are "mutually inverse". Before doing this, however, we wish to make some remarks on the group scheme Inv. $\underline{L}$.

In the classical theory, the group Inv. $\underline{L}$ is the Galois group of K/L. In the more general theory, this will be impossible since Aut (K/L) is a group scheme over Spec L, while Inv. L is a group scheme over Spec k. However, the next best thing is true, namely:

Proposition. The natural morphism

$$\text{Inv. } \underline{L} \otimes L \longrightarrow \underline{\text{Aut}}\ (K/L)$$

is an isomorphism for every subfield L of the layer K/k.

Suppose we start with a subgroupscheme H of Aut (K/k), and form the associated subfield

$$L = \text{Fix } \underline{H}\ (k)$$

of the layer K/k. Now we form the associated groupscheme Inv. L. What is the relation between H and Inv. L? Answer: There is a closed immersion $\underline{H} \hookrightarrow \text{Inv. } \underline{L}$.

Now by the usual "double commutant" reasoning, we can easily show that $L \subseteq \text{Fix (Inv. } \underline{L})\ (k)$. It is the converse inequality which requires more proof. However, if L is already the field associated to some groupscheme H, then one finds, upon application of the functor Fix to the closed immersion $\underline{H} \hookrightarrow \text{Inv. } \underline{L}$, that

$$L \subseteq \text{Fix (Inv. } \underline{L})\ (k) \subseteq \text{Fix } \underline{H}\ (k) = L.$$

Consequently: For those subfields L which arise as associated fields to a subgroupscheme H of Aut (K/k), we have: $L = \text{Fix (Inv. } \underline{L})\ (k)$. Moreover, in the Galois equivalence class corresponding to L, the groupscheme Inv. L is the unique maximal member.

3. GALOIS THEORY; COUNTING

We must now prove that for a given field extension K/k, every subfield L arises as associated field to a subgroupscheme of Aut (K/k). A necessary and sufficient condition that L be the "fixed field" of a subgroupscheme of Aut (K/k), is that

(*) $$L = \text{Fix}\ (\text{Inv.}\ \underline{L})\ (k)$$

The problem is connected with the existence of "sufficiently many" automorphisms as we shall now sketch. Because of the isomorphism

$$\text{Inv.}\ \underline{L} \otimes L \xrightarrow{\sim} \underline{\text{Aut}}\ (K/L),$$

equation (*) is clearly equivalent with

(**) $$k = \text{Fix}\ (\underline{\text{Aut}}\ (K/k))(k).$$

Equation (**) is, in turn, equivalent to the "moving property": Let ξ be an element of exponent one in K, then there exists a k-algebra A and an automorphism $\sigma \in \text{Aut}_A (K \otimes A)$ such that $\sigma(\xi) \neq \xi$.

If an extension K/k has the moving property, we shall say that it is a good extension. Now, in the oringinal form of this work, I proved that every extension had an embedding in a good extension by a fairly elaborate procedure. However, I have since shown that every purely inseparable layer K/k is good. Consequently, we may now summarize section 2 and state:

Theorem. Let K/k be a purely inseparable, finite, extension, then K/k is a good extension. The set of all closed subgroupschemes of Aut (K/k) is fibred by a natural equivalence relation; in each equivalence class there is a unique maximal member, and the correspondences

$$\mathrm{cl}\ (\underline{H}) \rightsquigarrow \mathrm{Fix}\ \underline{H}\ (k)$$

$$L \rightsquigarrow \mathrm{cl}\ (\mathrm{Inv}\ \underline{L})$$

are 1-1, lattice inverting, between the equivalence classes of subgroupschemes of Aut (K/k) and all the subfields of K/k. Moreover, Inv L is the unique maximal subgroupscheme in its class. (Actually, the statement of the theorem may be made more symmetric if one notes that the subringschemes of K/k are fibred by a natural equivalence relation. In each class there is a unique maximal member (a field!), and the correspondence sketched above actually runs between classes of subgroupschemes and classes of subringschemes.)

The subgroup theorem of Galois Theory has as a natural analog the statement: $\mathrm{Inv}\ \underline{L} \otimes L \xrightarrow{\sim} \underline{\mathrm{Aut}}\ (K/L)$. What about the famous connection between normal subfields and normal subgroups? It is here that striking differences occur between the general and classical cases.

Recall that if $\underline{H}$ is a subgroupscheme of $\underline{G}$, the normalizer of $\underline{H}$ in $\underline{G}$, denoted $N_{\underline{G}}\ \underline{H}$, is defined by the equation

$$N_{\underline{G}}\ \underline{H}\ (Ƶ) = \{\sigma \in \underline{G}\ (Ƶ) \mid \text{for every scheme } U \text{ over } Ƶ, \text{ the image of } \sigma \text{ in } \underline{G}\ (U) \text{ normalizes the subgroup } \underline{H}\ (U).\}.$$

Here $\mathfrak{Z}$ is a scheme over the same base as $\underline{G}$, and $\underline{G}(\mathfrak{Z})$, etc., means points of $\underline{G}$ with values in $\mathfrak{Z}$. The same sort of definition works for centralizers. A subgroupscheme $\underline{H}$ of $\underline{G}$ is normal in $\underline{G}$ when and only when $N_{\underline{G}} \underline{H} = \underline{G}$. We shall write $\underline{H} \triangleleft \underline{G}$ for this.

If K/k is a field extension, and if L is a subfield of the layer K/k, we shall say that L is K/k - normal, if and only if for every k-algebra A, every automorphism $\sigma \in \mathrm{Aut}_A(K\otimes A)$ transforms $\underline{L}(A)$ into itself. Observe that this is a relative concept - it depends upon the field K. One can easily give examples of a tower of fields, $k \subseteq K \subseteq \Omega$, where a subfield L of K/k is K/k-normal, but is NOT Ω/k-normal. The problem is that no longer does an automorphism map a given field element only on finitely many choices, it may have infinitely many. This makes normality very strongly dependent on the structure of $\underline{\mathrm{Aut}}$ (K/k).

Proposition. Let K/k be a field extension. A necessary and sufficient condition that a subfield L of K/k be K/k-normal is that Inv $\underline{L}$ be a normal subgroupscheme of $\underline{\mathrm{Aut}}$ (K/k). When K/k is a pure extension, (in the sense of Sweedler, [5]), the natural morphism (of group schemes)

$$\underline{\mathrm{Aut}}\,(K/k) \longrightarrow \underline{\mathrm{Aut}}\,(L/k)$$

given by restriction of an automorphism is a surjection of groupschemes with kernel Inv $\underline{L}$. Consequently,

$$\underline{\mathrm{Aut}}\,(L/k) \cong \underline{\mathrm{Aut}}\,(K/k)/\mathrm{Inv}\,\underline{L}.$$

Is such a large group as *Aut* (K/k) necessary? After all, the classical case gives us a finite group whose order is exactly the degree of the extension. Based on this, one would expect some sort of finite group scheme to work.

In the Jacobson Theory, a large role is played by the Lie Algebra of k-derivations of K, say L. But one knows that such an algebra is the tangent space of a uniquely determined finite group scheme of Frobenius height 1, say *G*, [4]. The dimension of L over K is n, where n is the number of generators necessary to obtain the field K over k. As [K: k] equals p^n, we find that $\dim_k L = n\,p^n$. Then one knows that the order of *G* is precisely $p^{\dim_k L} = (p^n)^{p^n} = [K:k]^{[K:k]}$. *Thus the group scheme associated to the extension* K/k *of exponent* 1 *by the Jacobson method has order* $[K:k]^{[K:k]}$.

Theorem. *Let* K/k *be a purely inseparable, pure extension in which each generator has the same exponent* r. *Let* *G* (K/k) *denote the kernel of the* r-*fold power of the Frobenius morphism on* *Aut* (K/k). *Then* *G* (K/k) *has order*

$$[K:k]^{\left(\frac{r+1}{2}\right)\sqrt[r]{[K:k]}}$$

and intersection of the classes of Galois equivalent subgroupschemes of *Aut* (K/k) *with* *G* (K/k) *yields a fibration of* *G* (K/k) *which still classifies the subfields of* K/k *according to the prescriptions above. When* r = 1, *G*(K/k) *is pre-*

<u>cisely the groupscheme yielded by the Jacobson method.</u>

<u>Remarks</u>

1. The classical theory can be put into this framework very easily once we realize that it is actually the structure of K ⊗ K as K-algebra which is at stake in the usual (and more general) Galois Theory. The normal basis theorem asserts that K ⊗ K is a direct product of copies of K indexed by the usual Galois group, and this group acts on K ⊗ K as K-automorphisms by left translation.

2. One can count Inv° <u>L</u> (the correct finite groupscheme corresponding to Inv <u>L</u>), and do the usual things with this counting process. A full exposition will be published when all the proofs of the above are published.

4. A CRITIQUE

The most distressing thing about the theory above is the use of classes of subgroupschemes. How can one characterize the maximal elements Inv <u>L</u> in each class? Probably, the groupschemes Inv <u>L</u> are exactly the K-stable groupschemes in each class, where K acts on <u>Aut</u> (K/k), or at least on the kernel of powers of Frobenius, <u>via</u> a generalized (perhaps even an Artin - Hasse) exponential.

I am endeavoring to answer this question in some satisfactory way.

REFERENCES

[1] Grothendieck, A., "Technique de descente et théorèmes d'existence en Géométrie Algébrique II", *Sém. Bourbaki*, [1959-1960], Éxp. 195.

[2] Grothendieck, A., and M. Demazure, *Séminaire Géométrie Algébrique de l' Institut des Hautes Études Scientifiques*, [1963-1964].

[3] Serre, J.P., "Quelques propriétés des variétés abéliennes en caractéristique p", *Amer. J. Math.*, 80; 715-739, [1958].

[4] Shatz, S.S., "Cohomology of Artinian groupschemes over local fields", *Ann. of Math*, 79; 411-449, [1964].

[5] Sweedler M., "Structure of purely inseparable extensions", *Ann. of Math.*, 87; 401-411, [1968].

DERIVED FUNCTORS WITHOUT INJECTIVES

by

H. B. Stauffer

1. INTRODUCTION

Given a covariant additive functor $F: \overline{A} \longrightarrow \overline{B}$ where $\overline{A}$ is a small abelian category and $\overline{B}$ is an abelian category satisfying AB 5, one can construct its right derived functors without using injectives. This can be accomplished by embedding $\overline{A}$ in a "larger" category $\overline{A}^{dir}$ which has injectives and allows a "unique" extension functor F^{dir} of F. The derived functors of F are obtained by taking the derived functors of F^{dir}, using the injectives in $\overline{A}^{dir}$, and restricting to $\overline{A}$. The category $\overline{A}^{dir}$ is actually a "right completion" of $\overline{A}$. I would like to thank Saul Lubkin for his help and encouragement.

2. $\overline{A}^{dir}$

Let $\overline{A}$ be any category. We shall construct a category $\overline{A}^{dir}$ from $\overline{A}$ in the following way. The objects of $\overline{A}^{dir}$ are directed systems $(A_i; \alpha_{ii'})_I$ in $\overline{A}$ over directed sets. The morphisms from $(A_i; \alpha_{ii'})_I$ to $(B_j; \beta_{jj'})_J$ are ordered pairs (ϕ, f) where $\phi: I \longrightarrow J$ is a set map and $f_i: A_i \longrightarrow B_{\phi(i)}$ is a morphism in $\overline{A}$

for each $i \in I$ such that $i' \geq i$ in I implies there exists $j \geq \phi(i), \phi(i')$ in J such that the diagram commutes:

$$\begin{array}{ccccc} A_{i'} & \xrightarrow{f_{i'}} & B_{\phi(i')} & \searrow^{\beta_{\phi(i')j}} & \\ \uparrow{\scriptstyle \alpha_{ii'}} & & & & B_j \\ A_i & \xrightarrow{f_i} & B_{\phi(i)} & \nearrow_{\beta_{\phi(i)j}} & \end{array} .$$

An equivalence relation is introduced. In $\mathrm{Hom}_{\overline{A}}\mathrm{dir}$ $((A_i;\ \alpha_{ii'})_I,\ (B_j;\ \beta_{jj'})_J)$ let $(\phi,f) \backsim (\phi',f')$ if and only if there exists (ϕ'',f'') dominating both; i.e., $\phi''(i) \geq \phi(i), \phi'(i)$ for all i and the diagrams commute:

$$\begin{array}{ccc} A_i & \xrightarrow{f''_i} & B_{\phi''(i)} \\ & {\scriptstyle f_i}\searrow & \uparrow{\scriptstyle \beta_{\phi(i)\phi''(i)}} \\ & & B_{\phi(i)} \end{array} \qquad \begin{array}{ccc} A_i & \xrightarrow{f''_i} & B_{\phi''(i)} \\ & {\scriptstyle f_{i'}}\searrow & \uparrow{\scriptstyle \beta_{\phi'(i)\phi''(i)}} \\ & & B_{\phi'(i)} \end{array} .$$

We shall refer to $(A_i;\ \alpha_{ii'})_I$ as $(A_i)_I$ where the meaning is clear.

<u>Proposition 2.1</u>

Let $(A^1)_{I_1}, (A^2)_{I_2}, \ldots, (A^n)_{I_n} \in \overline{A}^{\mathrm{dir}}$, and, for $1 \leq h \leq k \leq n$, let $M_{h,k} \subset \mathrm{Hom}_{\overline{A}}\mathrm{dir}\ ((A^h)_{I_h}, (A^k)_{I_k})$ be finite. Then there exists isomorphisms $\theta_r\colon (B^r)_I \xrightarrow{\simeq} (A^r)_{I_r}$ in $\overline{A}^{\mathrm{dir}}$, $1 \leq r \leq n$, and $M'_{h,k} = \theta_k^{-1} \circ M_{h,k} \circ \theta_h \subset \mathrm{Hom}_{\overline{A}}\mathrm{dir}\ ((B^h)_I, (B^k)_I)$ such that:

(i) all $(B^r)_I$ are indexed by the same directed set I, and

(ii) each morphism in $M'_{h,k}$ can be represented by (ϕ,f) where $\phi = 1_I$ and f is a map of directed systems.

Proof. If there are two objects $(A^1)_{I_1}$ and $(A^2)_{I_2}$ and no maps, we can construct directed set $I_1 \times I_2 = \{(i_1, i_2) | i_1 \in I_1, i_2 \in I_2, (i_1, i_2) \leq (i_1', i_2')$ if and only if $i_1 \leq i_1'$ and $i_2 \leq i_2'\}$. Then let $(B^r)_{I_1 \times I_2}$ be given by $B^r_{(i_1,i_2)} = A^r_{i_r}$, $r = 1,2$, with the obvious morphisms, and we are done.

If we have one map represented by $(\phi,f)\colon (A^1)_{I_1} \longrightarrow (A^2)_{I_2}$, then we can first choose $(B^1)_J \xrightarrow[\simeq]{\theta_1'} (A^1)_{I_1}$ and $(\phi',f') \sim (\phi,f)\circ\theta_1'$ whose image objects $\{A^2_{\phi'(j)} | j \in J\}$ in $(A^2)_{I_2}$ are indexed by a cofinal subset I_2. Then we can choose $(B^1)_I \xrightarrow[\simeq]{\theta_1''} (B^1)_J$ where each $i_o \in I$ is preceded by only a finite number of $i \in I$ (i.e., the set $\{i | i \leq i_o \text{ in } I\}$ is finite) and $(\phi'',f'') \sim (\phi',f')\circ\theta_1''$ whose image objects are indexed by a cofinal subset of I_2. Finally we can choose $(\phi''',f''') \sim (\phi'',f'')$ whose image objects give the desired $(B^2)_I$. We have our result:

$$\begin{array}{ccc} (A^1)_{I_1} & \xrightarrow{(\phi,f)} & (A^2)_{I_2} \\ \theta_1 = \theta_1'\circ\theta_1'' \uparrow \simeq & & \simeq \uparrow \theta_2 \\ (B^1)_I & \xrightarrow{(1,f''')} & (B^2)_I \end{array}$$

The situation with n maps can be handled by similar methods.

Theorem 2.2

Let $\overline{A}$ be any category. Then

(i) $\overline{A}^{dir}$ is closed under $\varinjlim$ (over directed sets).

(ii) The obvious covariant functor $I_{\overline{A}}: \overline{A} \longrightarrow \overline{A}^{dir}$ is a full embedding, and, given covariant functor $F: \overline{A} \longrightarrow \overline{B}$ where $\overline{B}$ has $\varinjlim$ (over directed sets), there exists a unique (up to equivalence) covariant functor $F^{dir}: \overline{A}^{dir} \longrightarrow \overline{B}$ which preserves $\varinjlim$ (over directed sets) and yields the commutative diagram:

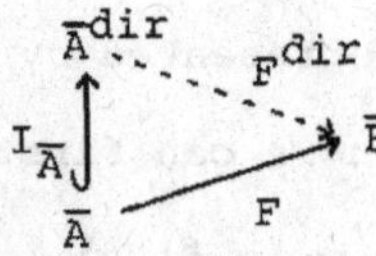

(iii) $\overline{A}$ additive implies $\overline{A}^{dir}$ is additive.

(iv) $\overline{A}$ small and abelian implies $\overline{A}^{dir}$ is abelian, satisfies AB 5, and has enough injectives. $I_{\overline{A}}: \overline{A} \longrightarrow \overline{A}^{dir}$ is exact.

Proof. In $\overline{A}^{dir}$ $\varinjlim_{j \varepsilon J} (A_i^j)_{i \varepsilon I_j} =$ $(A_i^j)_{i \varepsilon I_j,\ j \varepsilon J}$, the directed system obtained by the conglomerate of all objects and maps. F^{dir} is defined by $F^{dir}(A_i; \alpha_{ii'})_I = \varinjlim \{F(A_i); F(\alpha_{ii'})\}_I$ in $\overline{B}$. We can use Proposition 2.1 to show that $\overline{A}^{dir}$ is additive in (iii) and abelian in (iv); for example, kernels can then be computed "pointwise". The AB 5 property is immediate: $\varinjlim$ is exact. Since we have a set of generators in $\overline{A}^{dir}$, namely the objects of $\overline{A}$, there is a generator. This ensures the existence of enough injectives.

Let us say that $(A_i)_I \in \overline{A}^{dir}$ <u>admits a final object</u> $A \in \overline{A}$ if and only if $(A_i)_I$ is isomorphic in $\overline{A}^{dir}$ to (A).

<u>Proposition 2.3</u>

If $(A_i, \alpha_{ii'})_I$ admits a final object A, then

(i) there exists $i_o \in I$ such that, for each $j \geq i_o$, $A_j = A \oplus B_j$ for some $B_j \in \overline{A}$, and

(ii) for each $j \geq i_o$, there exists $j' \geq j$ such that $\alpha_{jj'} = 1 \oplus 0\colon A \oplus B_j \longrightarrow A \oplus B_{j'}$.

<u>Proof</u>. If $(A_i, \alpha_{ii'})_I$ admits a final object, then we have isomorphisms $(\phi,f)\colon (A) \longrightarrow (A_i, \alpha_{ii'})_I$ and its inverse $(\rho,g)\colon (A_i, \alpha_{ii'})_I \longrightarrow (A)$. We let i_o be the index of the image object of (ϕ,f) and, looking at the composition of the isomorphisms, easily obtain the results.

3. <u>DERIVED FUNCTORS</u>

Let F, $\{F^n\}_{n\geq 0} = F^*\colon \overline{A} \longrightarrow \overline{B}$ be covariant additive functors with $\overline{A}$ small and abelian, $\overline{B}$ abelian satisfying AB 5, and F^* a connected sequence of functors. Let $\eta\colon F \longrightarrow F^o$ be a natural transformation.

Let $\{A^*: 0 \longrightarrow A \longrightarrow A^o \longrightarrow A' \longrightarrow \cdots \mid A^*$ is a resolution of A in $\bar{A}\}$ be the objects of the small category $\bar{C}_A$ with morphisms all maps of resolutions of A lifting 1_A in $\bar{A}$. Given $f: A \longrightarrow B$ and resolution A^* of A, by using pushouts we get a resolution $f^*(A^*)$ of B; $f^*: \bar{C}_A \longrightarrow \bar{C}_B$ is a functor. We then have the induced functors $S^n: \bar{A} \longrightarrow \bar{B}$, $n \geq 0$, given by $S^n(A) = \underset{A^* \varepsilon \bar{C}_A}{\overrightarrow{\lim}} H^n(F(A^*))$.

<u>Theorem 3.1</u>

The following are equivalent:

(i) $\eta: F \longrightarrow F^o$ is universal left exact (i.e., given natural transformation $\xi: F \longrightarrow L$, a covariant left exact functor, there exists a unique natural transformation $\tau: F^o \longrightarrow L$ such that the diagram

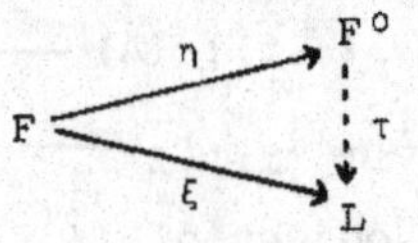

commutes), and F^{n+1} is the right satellite of F^n, $n \geq 0$;

(ii) The connected sequence of functors F^* is exact, and, if $G^* = \{G^n\}_{n \geq 0}$ is an exact connected sequence of functors with $\gamma: F \longrightarrow G^o$ a natural transformation, then there exists unique natural transformations $\{f^n\}_{n \geq 0} = f^*: F^* \longrightarrow G^*$ such that the diagram

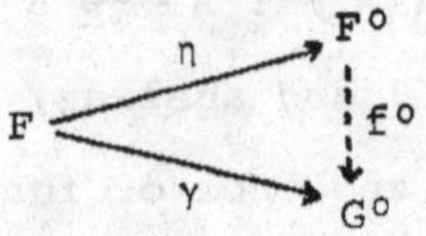

commutes;

(iii) Consider the diagram

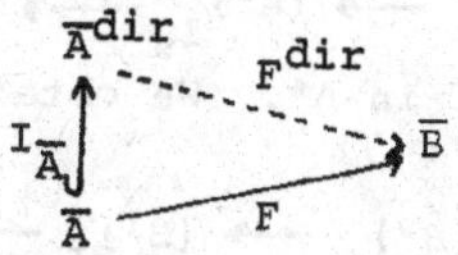

Then $F^n = R^n(F^{dir})\Big|_{\overline{A}} = R^n(F^{dir}) \circ I_{\overline{A}}$, $n \geq 0$, where the $R^n(F^{dir})$ are the right derived functors of F^{dir} (defined on a category with enough injectives);

(iv) $F^n = S^n$, $n \geq 0$.

Proof. An (exact) connected sequence $F^* = \{F^n\}_{n\geq 0}: \overline{A} \longrightarrow \overline{B}$ lifts to an (exact) connected sequence $F^{*dir} = \{(F^n)^{dir}\}_{n\geq 0}: \overline{A}^{dir} \longrightarrow \overline{B}$. Similarly a left exact functor $F: \overline{A} \longrightarrow \overline{B}$ lifts to a left exact functor $F^{dir}: \overline{A}^{dir} \longrightarrow \overline{B}$. A natural transformation $\eta: F \longrightarrow G$ lifts to a natural transformation $\eta^{dir}: F^{dir} \longrightarrow G^{dir}$ defined by $\eta^{dir}(A_i)_I = \varinjlim_I \{\eta(A_i)\}$ in $\overline{B}$. Using these observations, (i) $\Longleftrightarrow$ (iii) and (ii) $\Longrightarrow$ (iii) are straightforward. We conclude by proving (iii) $\Longleftrightarrow$ (iv). If we let $\{A^*: 0 \longrightarrow (A) \longrightarrow (A^o)_{I_o} \longrightarrow (A^1)_{I_1} \longrightarrow \dots \mid A^*$ is a resolution of (A) in $\overline{A}^{dir}\}$ be the objects of the category $\overline{C}_A^{dir}$ with morphisms all maps of resolutions of (A) lifting $1_{(A)}$ in $\overline{A}^{dir}$, we have the induced functors $T^n: \overline{A}^{dir} \longrightarrow \overline{B}$, $n \geq 0$, given by $T^n(A) = \varinjlim_{A^* \epsilon \overline{C}_A^{dir}} H^n(F^{dir}(A^*))$. It is immediate that $R^n(F^{dir}) = T^n$, $n \geq 0$. Clearly $\overline{C}_A \subset \overline{C}_A^{dir}$.

Conversely, if $A^* \varepsilon \bar{C}_A^{dir}$, we can use Proposition 2.1 on the finite maps $0 \longrightarrow (A) \longrightarrow (A^o)_{I_o} \longrightarrow (A')_{I_1} \longrightarrow \cdots \longrightarrow (A^n)_{I_n} \longrightarrow (A^{n+1})_{I_{n+1}}$ in A*. We obtain the isomorphic sequence $0 \longrightarrow (B)_I \longrightarrow (B^o)_I \longrightarrow (B^1)_I \longrightarrow \cdots \longrightarrow (B^n)_I \longrightarrow (B^{n+1})_I$. Since $H^n(F^{dir}(A^*)) =$

$H^n(\varinjlim_I F(B_i^n)_I) = \varinjlim_I H^n(F(B_i^n))$, we may use Proposition 2.3 on $(B)_I$. The situation reduces to looking at resolutions over $A \oplus B_j$, $j \geq i_o$ in I. It is easy to see that such resolutions map into resolutions over A.
Hence, we are done.

REFERENCES

[1] Buchsbaum, D. A., "Satellites and Universal Functors", *Ann. of Math.*, 71; 199-209, (1960).

[2] Freyd, P. J., *Abelian Categories*, Harper and Row: New York, 1964.

[3] MacLane, S., *Homology*, Springer: Berlin, 1963.

[4] Mitchell, B., *Theory of Categories*, Academic Press: New York, 1965.

SIMPLICIAL DERIVED FUNCTORS

by

Myles Tierney and Wolfgang Vogel

1. INTRODUCTION

In this note we sketch (details will appear elsewhere) a theory of derived functors that is a simple generalization of the classical procedure, except that kernels are replaced by "simplicial kernels", and absolute projectives are replaced by relative ones. With this theory one can derive arbitrary functors $E: \underline{A} \longrightarrow \underline{B}$ if $\underline{A}$ has finite limits and a projective class and $\underline{B}$ is abelian.

After introducing the basic concepts and stating the comparison theorem, the derived functors are defined in §2. In §3 we show how to compare the theory of §2 with other notions of derived functors. In particular, if there is a cotriple $\mathfrak{G}$ in $\underline{A}$ that realizes the given projective class, then we obtain the cotriple derived functors of Barr and Beck [2]; if we choose the projective class as models, then the derived functors of §2 agree with those of André [1]. Also, if $\underline{A}$ is abelian then our procedure yields the (suitably relativized) 0-level derived functors of Dold-Puppe [3], and if in addition E is additive, then we obtain the relative theory of Eilenberg and Moore [4].

We remark that the method of §2 was also used by Evrard [5] in the absolute case. Michael Barr has informed us that both he and Max Kelley have also considered this theory (unpublished).

2. DERIVED FUNCTORS

2.1 Definition

Let $\underline{A}$ be a category, and

$$X \underset{f^n}{\overset{f^\circ}{\rightrightarrows}} Y \quad (\text{arrows } f^\circ, \ldots, f^n)$$

a sequence of $n + 1$ $\underline{A}$-morphisms for $n \geq 0$. A **simplicial kernel** of $(f^\circ, ---, f^n)$ is a sequence

$$K \underset{k^{n+1}}{\overset{k^\circ}{\rightrightarrows}} X \quad (\text{arrows } k^\circ, \ldots, k^{n+1})$$

of $n + 2$ $\underline{A}$-morphisms satisfying $f^i k^j = f^{j-1} k^i$ for $0 \leq i < j \leq n + 1$, and universal with respect to this property. That is, if

$$H \underset{h^{n+1}}{\overset{h^\circ}{\rightrightarrows}} X \quad (\text{arrows } h^\circ, \ldots, h^{n+1})$$

is any other sequence satisfying $f^i h^j = f^{j-1} h^i$ for $0 \leq i < j \leq n+1$, then there exists a unique $h: H \longrightarrow K$ such that $k^i h = h^i \quad 0 \leq i \leq n+1$.

Obvious dualization yields the notion of a co-simplicial kernel, and later a theory of derived functors with respect to an injective class.

Clearly, simplicial kernels are unique up to isomorphisms if they exist, and concerning existence one can prove easily:

<u>2.2 Proposition</u>

If $\underline{A}$ has finite limits, then simplicial kernels exist for any sequence $(f^\circ, ---, f^n)$ and any $n \geq 0$.

We review briefly the notion of a projective class in $\underline{A}$, for details see [4]. Let P be a class of objects of $\underline{A}$. We say that $f: A \longrightarrow A'$ is <u>P-epimorphic</u> if for all $X \in P$,

$$\underline{A}(X,f) : \underline{A}(X,A) \longrightarrow \underline{A}(X,A')$$

is surjective. We call P a <u>projective class</u> if for each $A \in \underline{A}$ there exists a P-epimorphism $e: X \longrightarrow A$ with $X \in P$.

A <u>simplicial object of $\underline{A}$ augmented over</u> A (without degeneracies) in a diagram

$$\cdots X_n \underset{\partial_n^n}{\overset{\partial_n^o}{\rightrightarrows}} X_{n-1} \underset{\partial_{n-1}^{n-1}}{\overset{\partial_{n-1}^o}{\rightrightarrows}} X_{n-2} \cdots X_1 \underset{\partial_1^1}{\overset{\partial_1^o}{\rightrightarrows}} X_o \xrightarrow{\partial_o^o} A$$

in $\underline{A}$, written $X \xrightarrow{\partial^o} A$, such that

$$\partial_{n-1}^i \, \partial_n^j = \partial_{n-1}^{j-1} \, \partial_n^i \qquad 0 \leq i < j \leq n \quad .$$

The ∂_n^i are called <u>face operators</u>, and in the future we will denote a diagram such as

$$X_n \underset{\partial_n^n}{\overset{\partial_n^o}{\rightrightarrows}} X_{n-1}$$

simply by $X_n \xrightarrow{\partial_n^i} X_{n-1}$, agreeing always that i runs from 0 to n. We will also often omit the subscripts in morphisms such as ∂_n^i if this does not lead to confusion. If $f\colon A \longrightarrow A'$, and $X \xrightarrow{\partial^o} A$ and $X' \xrightarrow{\partial^o} A'$ are augmented simplicial objects over A and A', then a <u>morphism</u> $\bar{f}$ <u>over</u> f, written

$$\begin{array}{ccc} X & \xrightarrow{\bar{f}} & X' \\ {\scriptstyle \partial^o}\downarrow & & \downarrow{\scriptstyle \partial^o} \\ A & \xrightarrow{f} & A' \end{array}$$

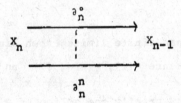

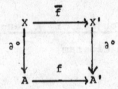

consists of $\underline{A}$-morphisms $\bar{f}_n: X_n \longrightarrow X'_n$ for $n \geq 0$ satisfying

$$\partial^i_n \bar{f}_n = \bar{f}_{n-1} \partial^i_n$$

for $0 \leq i \leq n$, where we put $f_{-1} = f$.

If P is a projective class in $\underline{A}$, and $X \xrightarrow{\partial^\circ} A$ is an augmented simplicial object over A, then we say $X \xrightarrow{\partial^\circ} A$ is P-projective if each $X_n \in P$. If $\underline{A}$ has simplicial kernels, then we have a factorization

(2.3)

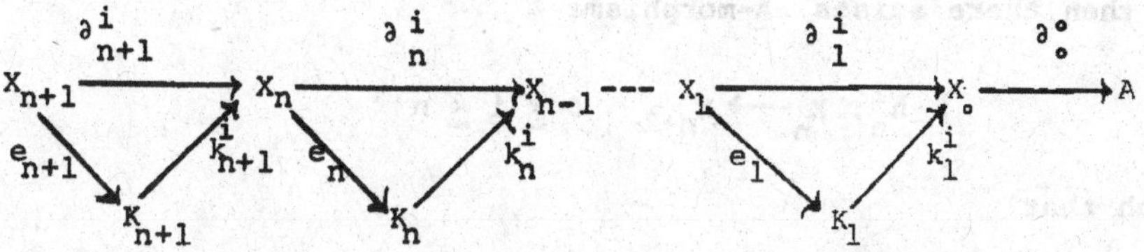

where $K_n \xrightarrow{k^i_n} X_{n-1}$ is a simplicial kernel of $X_{n-1} \xrightarrow{\partial^i_{n-1}} X_{n-2}$ for $n \geq 1$, if we put $X_{-1} = A$. $X \xrightarrow{\partial^\circ} A$ is said to be P-<u>exact</u> if $\partial^\circ_\circ$ and e_n $n \geq 1$ are P-epimorphic. $X \xrightarrow{\partial^\circ} A$ is a P-<u>projective resolution</u> of A if it is P-projective and P-exact. By 2.2, if $\underline{A}$ has finite limits then each $A \in \underline{A}$ can be so resolved. Moreover, such a resolution is unique up to simplicial homotopy equivalence as results immediately from the following comparison theorem.

2.4 Theorem

Let $X \xrightarrow{\partial^\circ} A$ be a P-projective and $X' \xrightarrow{\partial^\circ} A'$ be P-exact. Then any morphism $f: A \longrightarrow A'$ can be extended to a morphism

$$\begin{array}{ccc} X & \xrightarrow{\bar{f}} & X' \\ {\scriptstyle \partial^\circ}\downarrow & & \downarrow{\scriptstyle \partial^\circ} \\ A & \xrightarrow{f} & A' \end{array}$$

over f. Furthermore, any two such extensions are simplicially homotopic. That is, if $\bar{f}, \bar{g}: X \longrightarrow X'$ are two extensions of f, then there exists A-morphisms

$$h^i_n: X_n \longrightarrow X_{n+1} \qquad 0 \leq i \leq n$$

such that

$$\partial^\circ_{n+1}\, h^\circ_n = \bar{f}_n$$

$$\partial^{n+1}_{n+1}\, h^n_n = \bar{g}_n$$

$$\partial^i\, h^j = \begin{cases} h^{j-1}\, \partial^i & i < j \\ h^j\, \partial^{i-1} & i > j+1 \end{cases}$$

and $$\partial^{j+1}\, h^{j+1} = \partial^{j+1}\, h^j \quad .$$

By using similar (but easier) techniques, one can show that a P-projective resolution $X \xrightarrow{\partial^\circ} A$ has <u>pseudo-degeneracy operators</u>. That is, there exist

$$s^j_n : X_n \longrightarrow Y_{n+1} \qquad 0 \leq j \leq n$$

such that

$$\partial^i s^j \quad \begin{cases} s^{j-1}\,\partial^i & i < j \\ & i = j,\ j+1 \\ s^j\,\partial^{i-1} & i > j+1 \end{cases} .$$

In general the identity $s^i s^j = s^{j+1} s^i \quad i \leq j$ is not satisfied, but, as we shall see, can be achieved when $\underline{A}$ is abelian.

Given 2.4, it is clear how to derive a functor $E: \underline{A} \longrightarrow \underline{B}$ when $\underline{B}$ is abelian and $\underline{A}$ has finite limits and a projective class P. First, if Y is a simplicial object (with or without degeneracies) in an additive category, let us denote by kY the chain complex whose component in dimension n is Y_n, and whose boundary operator $\partial_n : Y_n \longrightarrow Y_{n-1}$ is $\sum_{i=0}^{n} (-1)^i\, \partial^i_n$. Now if $A \in \underline{A}$ choose a P-projective resolution $X \xrightarrow{\partial^\circ} A$ of A and define

$$L_n E: \underline{A} \longrightarrow \underline{B}$$

for $n \geq 0$ by setting

$$L_n E(A) = H_n(k\,E\,X) .$$

The definition is independent (up to isomorphism) of the resolution chosen and functorial in A by 2.4.

3. COMPARISON WITH OTHER THEORIES

Let $\underline{A}$ be a category with finite limits and a projective class P. We have

3.1 Lemma

Let $A \in \underline{A}$. If $Y \xrightarrow{\partial^0} A$ is P-exact, then for each $X \in P$ the augmented simplicial set (without degeneracies)

$$\underline{A}(X,Y) \longrightarrow \underline{A}(X,A)$$

has a contraction. That is, if we set $Y_{-1} = A$, then there are functions

$$h_n : \underline{A}(X,Y_n) \longrightarrow \underline{A}(X,Y_{n+1})$$

for $n \geq -1$ such that if $f: X \longrightarrow Y_n$, then

$$\partial^i_{n+1} h_n (f) = h_{n-1} (\partial^i_n f) \qquad 0 \leq i \leq n$$

and

$$\partial^{n+1}_{n+1} h_n (f) = f \qquad .$$

Now if $E: \underline{A} \longrightarrow \underline{B}$ is an arbitrary functor where $\underline{B}$ is abelian, choose as models for the André theory [1] the full subcategory of $\underline{A}$ whose objects are the projectives in P. Denoting the resulting homology theory by $H_n(\ ,E)$, we have by 3.1 and 4.7 of [1],

3.2 Proposition

There are natural (in A and E) isomorphisms

$$L_n E \xrightarrow{\approx} H_n(\ ,E) \quad .$$

Let $G = (\mathbb{G},\varepsilon,\delta)$ be a cotriple in $\underline{A}$ such that $P = P_{\mathbb{G}}$, where $P_{\mathbb{G}}$ is the class of $\mathbb{G}$-projectives [2]. The existence of the contraction given in 3.1 implies easily that any P-projective resolution in the sense of §2 is a simplicial $\mathbb{G}$-resolution in the sense of [2], §5.1, and it is shown there that cotriple homology may be computed with such resolutions. (Simplicial $\mathbb{G}$-resolutions are assumed to have degeneracies in [2], but these play no role in the argument.) Hence we obtain

3.3 Proposition

If $\mathbb{G}$ is a cotriple in $\underline{A}$ with $P = P_{\mathbb{G}}$, then there are natural isomorphisms

$$L_n E \xrightarrow{\approx} H_n(\ ,E)_{\mathbb{G}} \qquad n \geq 0 \quad .$$

Since most classical cohomology theories appear as examples of cotriple cohomology, 3.3 provides a large class of easily constructed simplicial resolutions from which to compute these theories.

Assume now that $\underline{A}$ is abelian. Since $\underline{A}$ has an 0-object and pullbacks, it has all finite limits. If X is a simplicial object in $\underline{A}$ (with or without degeneracies) then the <u>normal</u> <u>complex</u> NX of X is the chain complex given by

$$(NX)_0 = X_0$$

$$(NX)_n = \bigcap_{i>0} \ker \{\partial^i : X_n \longrightarrow X_{n-1}\} \quad n > 0$$

and $\qquad \partial_n = \partial^\circ_n | (NX)_n \quad .$

If P is a projective class in $\underline{A}$, we have

4.1 Proposition

If the augmented simplicial object $X \xrightarrow{\partial^\circ} A$ is P-exact in the sense of §2, then the augmented chain complex $NX \xrightarrow{\partial^\circ} A$ is P-exact in the sense of Eilenberg-Moore [4], i.e. for each $P \in P$ the sequence of abelian groups

$$\ldots \longrightarrow \underline{A}(P, (NX)_n) \longrightarrow \underline{A}(P, (NX)_{n-1}) \longrightarrow \ldots$$

$$\longrightarrow \underline{A}(P, X_\circ) \longrightarrow \underline{A}(P, A) \longrightarrow 0$$

is exact.

Let us denote by $S(\underline{A})$ the category of genuine simplicial objects over $\underline{A}$. That is, an $X \in S(\underline{A})$ has faces and degeneracies and all the usual simplicial identities are satisfied. Morphisms commute with faces and degeneracies. We write $C(\underline{A})$ for the category of positive chain complexes over $\underline{A}$. Then, the normal complex is a functor $N: S(A) \longrightarrow C(A)$. In [3] it is shown that N has an inverse, i.e. there is a functor $K: C(\underline{A}) \longrightarrow S(\underline{A})$ and for $C \varepsilon C(\underline{A})$ and $X \varepsilon S(\underline{A})$ there are natural equivalences

$$\phi C: C \xrightarrow{\approx} NKC$$

$$\psi X: KNX \xrightarrow{\approx} X$$

It follows from the construction of K given in [3] that if we assume (as we now do) that P is closed under the formation of coproducts and retracts, then $(KC)_n \varepsilon P$ for each $n \geq 0$ iff $C_n \varepsilon P$ for each $n \geq 0$. Now if X has only pseudo-degeneracy operators, we can still, by choosing a fixed order in which to write iterated degeneracies, define a

$$\psi X: KNX \longrightarrow X$$

that commutes with all face operators and the <u>last</u> pseudo-degeneracy operator in each dimension. This ψ is natural with respect to morphisms $f: X \longrightarrow Y$ commuting with faces and degeneracies, where Y is also assumed to have only pseudo-degeneracies. If X has genuine degeneracies, meaning they satisfy $s^i s^j = s^{j+1} s^i$ $i \leq j$, then this ψ is the above ψ of Dold-Puppe, and then, of course, it commutes with all degeneracies. In any case, commutation with faces and the last degeneracy is enough to prove Lemma 3.17 of [3], which when applied to ψX shows that $(\psi X)_n$ is an isomorphism for $n \geq 0$.

In essence, what we have proved is that the pseudo-degeneracies of X can be replaced by degeneracies satisfying all the simplicial identities. A similar fact has been proved by Barr (unpublished) by means of a direct, inductive approximation process.

Let $E: \underline{A} \longrightarrow \underline{B}$ be an arbitrary functor. Then the (relativized) method of Dold-Puppe for defining (0-level) derived functors is the following. For $A \in \underline{A}$ choose a P-projective resolution $C \xrightarrow{\partial} A$ in the sense of Eilenberg-Moore. That is, $C \xrightarrow{\partial} A$ is P-projective and P-exact as defined in 4.1. Then set

$$L'_n E(A) = H_n(kEKC) \quad .$$

By the method of §2, choose a P-projective simplicial resolution $X \xrightarrow{\partial} A$ and assume it has pseudo-degeneracies. The derived functors of §2 are given by

$$L_n E(A) = H_n(kEX) \quad ,$$

but we have

$$(\psi X)_n : (KNX)_n \xrightarrow[\approx]{} X_n$$

for each $n \geq 0$. Hence, since P is closed it follows by 4.1 that $NX \xrightarrow{\partial} A$ is an Eilenberg-Moore P-projective resolution of A. Since ψX induces an isomorphism

$$k(EKNX) \xrightarrow[\approx]{} k(EX)$$

of chain complexes in $\underline{B}$, we obtain isomorphisms

(4.2) $$L'_n E(A) \xrightarrow[\approx]{} L_n E(A) \qquad n \geq 0 \quad .$$

Here a little care must be taken to show that these isomorphisms are natural in A.

Finally, consider the case $E: \underline{A} \longrightarrow \underline{B}$ where $\underline{A}$ and $\underline{B}$ are abelian and E is additive. If $A \in \underline{A}$ and $X \xrightarrow{\partial^\circ} A$ is a $\mathcal{P}$-projective resolution in the sense of §2, then

$$kEX = EkX$$

as chain complexes in $\underline{B}$. Letting $\sim$ and $\approx$ denote, respectively, chain equivalence and isomorphism of chain complexes, we have

(4.4) $$kX \approx k(KNX) \sim N(KNX) \approx NX \quad .$$

Since $NX \xrightarrow{\partial^\circ} A$ is a $\mathcal{P}$-projective resolution in the sense of Eilenberg-Moore, it follows from (4.3) that $kX \xrightarrow{\partial^\circ} A$ is also. Thus the derived functors of §2 coincide with those of [4]. Naturality is trivial in this case.

REFERENCES

[1] M. Andre, "Méthode Simpliciale en Algèbre Homologique et Algèbre Commutative", *Springer Lecture Notes in Mathematics*, 32, (1967).

[2] M. Barr, J. Beck, "Homology and Standard Constructions," (to appear in *Springer Lecture Notes*.)

[3] A. Dold, D. Puppe, "Homologie nicht-additiver Functoren". *Ann. Inst. Fourier*, 11; 201-312, (1961).

[4] S. Eilenberg, J. Moore, "Foundations of Relative Homological Algebra", *AMS Memoir*, 55, (1965).

[5] M. Evrard, "Homologie dans les Catégories non Abéliennes", *C.R. Acad. Sc. Paris*, 260; 749-751, 1044-1051, (1965).

ACYCLIC MODELS AND KAN EXTENSIONS

by Friedrich Ulmer

Introduction

The aim of this note is to point out the relationship between the technique of acyclic models and the standard procedure in homological algebra to compute derived functors by means of projectives.[1] This leads to a useful generalization of acyclic models reflecting the fact that the derived functors of a functor G can be computed more generally by G-acyclic resolutions.

Let $\underline{C}$ and $\underline{A}$ be categories, $\underline{A}$ abelian with sums and enough projectives, and let $J : \underline{M} \to \underline{C}$ be the inclusion of a full small category, the objects of which are referred to as models. Then the functor category $[\underline{M},\underline{A}]$ has also enough projectives and the restriction $R_J : [\underline{C},\underline{A}] \to [\underline{M},\underline{A}]$, $T \rightsquigarrow T\cdot J$, has a left adjoint $E_J : [\underline{M},\underline{A}] \to [\underline{C},\underline{A}]$ called the Kan extension. Note that for every $t : \underline{M} \to \underline{A}$ the functor $E_J(t) : \underline{C} \to \underline{A}$ is an extension of t . Roughly speaking, the technique of acyclic models turns out to be the standard procedure in homological algebra to compute the left derived functors of the Kan extension $E_J : [\underline{M},\underline{A}] \to [\underline{C},\underline{A}]$ by means of projective resolu-

1) The method of acyclic models was introduced by Eilenberg-MacLane [6]. Barr-Beck [2] gave a different version by means of cotriples.

tions.[2)]

Eilenberg-MacLane [6] showed that two complexes of functors T_* , $\bar{T}_*$: $\underline{C} \to \underline{A}$ with isomorphic augmentations $T_{-1} \to \bar{T}_{-1}$ are homotopically equivalent, provided the functors T_n , $\bar{T}_n$ are "representable" for $n \geq 0$ and the complexes $T_*M \to T_{-1}M \to 0$ and $\bar{T}_*M \to \bar{T}_{-1}M \to 0$ are exact for every model $M \in \underline{M}$. We show that a functor $T : \underline{C} \to \underline{A}$ is "representable" iff the restriction $T \cdot J$ is projective in $[\underline{M},\underline{A}]$ and the Kan extension of $T \cdot J$ is again T (i.e. $E_J(T \cdot J) = T$). Thus the above homotopy equivalence between T_* and $\bar{T}_*$ can be obtained in the following way [3)]; The complexes $T_* \cdot J$ and $\bar{T}_* \cdot J$ are projective resolutions of $T_{-1} \cdot J$ and $\bar{T}_{-1} \cdot J$. Hence there is a homotopy equivalence $h_* : T_* \cdot J \simeq \bar{T}_* \cdot J$ which is compatible with the augmentation $T_{-1} \cdot J \cong \bar{T}_{-1} \cdot J$. Applying the Kan extension $E_J : [\underline{M},\underline{A}] \to [\underline{C},\underline{A}]$ yields a homotopy equivalence

---

2) I observed this after I had received a first draft of a paper by Barr-Beck in 1967. Their improved version of acyclic models in [3] §XI and the fact that cotriple derived functors are the left derived functors of the Kan extension make fairly obvious that acyclic models and Kan extensions are closely related. In a recent paper of Dold, MacLane, Oberst [5] a relationship between acyclic models and projective classes was pointed out. As in André [1] and Barr-Beck [3] the Kan extension does not enter into the picture of [5] and all considerations are carried out in $[\underline{C},\underline{A}]$, the range of the Kan extension. It appears that the use of the Kan extension establishes a much closer relationship between acyclic models and homological algebra than the one in [5]. Moreover, it gives rise to a useful generalization of acyclic models which cannot be obtained by the methods of [5].

3) This was also observed by D. Swan (unpublished).

$E_J(h_*) : T_* \simeq \bar{T}_*$ because by the above $E_J(T_* \cdot J) = T_*$ and $E_J(\bar{T}_* \cdot J) = \bar{T}_*$ hold.

This procedure also shows that the n-th homology of T_* is the value of the n-th left derived functor of E_J at $T_{-1} \cdot J$, i.e. $L_n E_J(T_{-1} \cdot J) \cong H_n(T_*)$ holds. Moreover, one can show that $H_n(T_*) : \underline{C} \to \underline{A}$ is isomorphic with the n-th homology functor $H_n(-, T_{-1} \cdot J) : \underline{C} \to \underline{A}$ of André [1] p.3, provided $\underline{A}$ is a Grothendieck AB4) category. (André denotes the category $\underline{C}$ by $\underline{N}$.) It is well known that the left derived functors of E_J can be computed more generally by E_J-acyclic resolutions. This suggests defining a functor $T : \underline{C} \to \underline{A}$ as "weakly representable" if $E_J(T \cdot J) = T$ and $L_n E_J(T \cdot J) = 0$ for $n > 0$. This weaker notion of "representability", together with acyclicity, no longer guarantees a homotopy equivalence between the complexes T_* and $\bar{T}_*$ as above, but only that they have isomorphic homology and that every existing chain map $f_* : T_* \to \bar{T}_*$ compatible with the augmentation induces a homology isomorphism. Moreover, every chain map on the models $T_* \cdot J \to \bar{T}_* \cdot J$ can be extended to a chain map $T_* \to \bar{T}_*$ by means of the Kan extension $E_J : [\underline{M}, \underline{A}] \to [\underline{C}, \underline{A}]$. The lack of a homotopy equivalence can be compensated by conditions on additive functors $F : \underline{A} \to \underline{A}'$, guaranteeing that $H_n(F \cdot T_*) \cong H_n(F \cdot \bar{T}_*)$ is still valid and that $Ff_* : F \cdot T_* \to F\bar{T}_*$ induces a homology isomorphism.

Morphism sets, natural transformation and functor categories are denoted by brackets $[-,-]$, comma categories by parentheses (-,-) . We assume that all our categories are U-categories, where U is a sufficiently large universe, and we use the same terminology as Verdier in the appendix of "Sémi-naire geometrie algebrique 1964". We mostly refrain from mentioning this universe except when we also consider categories belonging to a larger universe. The categories of sets and abelian groups are denoted by $\underline{S}$ and $\underline{Ab}.\underline{Gr}$. The phrase "Let $\underline{A}$ be a category with direct limits" always means that $\underline{A}$ has direct limits over U-small index categories. However, we sometimes also consider direct limits of functors $F : \underline{D} \to \underline{A}$, where $\underline{D}$ is not necessarily small. Of course we then have to prove that this specific limit exists.

Our approach is based on the notions of generalized representable functors (cf. Ulmer [11]) and Kan extensions, (cf. Kan [8], Ulmer [13]) which prove very useful in this context. We first recall these notions and some of their properties.

(1) Let $\underline{M}$ and $\underline{A}$ be categories, $\underline{A}$ with sums. To each pair of objects $A \in \underline{A}$ and $M \in \underline{M}$ there is associated a generalized representable functor $\underline{M} \to \underline{A}$ (cf. [11]). It is the composite

$$A \otimes [M,-] : \underline{M} \to \underline{S} \to \underline{A}$$

where $[M,-] : \underline{M} \to \underline{S}$ is the hom-functor and $A \otimes : \underline{S} \to \underline{A}$ is the left adjoint of $[A,-] : \underline{A} \to \underline{S}$. Recall that $A\otimes : \underline{S} \to \underline{A}$ assigns to a set Λ the Λ-fold sum of A .[4)]

(2) Yoneda lemma, cf. [11]. Let $t : \underline{M} \to \underline{A}$ be a functor. Then there is a bijection

$$[A\otimes[M,-],t] \cong [A,tM]$$

natural in A , M and t . If $\underline{A}$ is additive, the bijection is an isomorphism of abelian groups.

Proof. The functors $A\otimes : \underline{S} \to \underline{A}$ and $[A,-] : \underline{A} \to \underline{S}$ induce a pair of adjoint functors $[\underline{M},\underline{S}] \to [\underline{M},\underline{A}]$, $s \rightsquigarrow A\otimes \cdot s$, and $[\underline{M},\underline{A}] \to [\underline{M},\underline{S}]$, $r \rightsquigarrow [A,-]\cdot r$. Thus we obtain by adjointness and the usual Yoneda lemma (for $r = t$ and $s = [M,-]$).

$$[A\otimes[M,-],t] \cong [[M,-] , [A,-]\cdot t] \cong [A,tM]$$

Q.E.D.

(3) Lemma. Let $t, t' : \underline{M} \to \underline{A}$ be functors and $\psi : t \to t'$ a natural transformation which is objectwise split, i.e., for each $M \in \underline{M}$ there is a morphism $\sigma(M) : t'M \to tM$ such that $\psi(M)\cdot\sigma(M) = \mathrm{id}$. (Note that σ need not be a natural transformation.) Then every diagram

---

4) For the dual notion of a corepresentable functor $\underline{M} \to \underline{A}$ we refer to Ulmer [11]. Note that a corepresentable functor is also covariant. The relationship between generalized representable and corepresentable functors is entirely different from the relationship between covariant and contravariant hom-functors.

$$\begin{array}{ccc} & & t \\ & \nearrow & \downarrow \psi \\ A\otimes[M,-] & \longrightarrow & t' \end{array}$$

can be completed as indicated. In particular, if $\underline{A}$ is abelian, this shows that the generalized representable functors are projective relative to the class $\mathcal{F}$ of short exact sequences in $[\underline{M},\underline{A}]$ which are objectwise split exact.

Proof. Since $\psi(M) : tM \to t'M$ is split, the induced map $[A,tM] \to [A,t'M]$ is epimorphic. Thus the assertion follows from the Yoneda lemma (2).

(4) Remark. It is obvious that (3) remains valid if $A\otimes[M,-]$ is replaced by a direct sum $\bigoplus_i(A_i\otimes[M_i,-])$.

(5) Another variation of (3) is the following: If $\underline{A}$ is abelian and $A \in \underline{A}$ projective, then the epimorphism $\psi : t \to t'$ need not be objectwise split, because the hom-functor $[A,-]$ preserves epimorphisms. This shows that $A\otimes[M,-]$ and likewise $\bigoplus_i(A_i\otimes[M_i,-])$ are projectives in $[\underline{M},\underline{A}]$, provided the objects A_i and A are projective in $\underline{A}$. Using constant functors $\underline{M} \to \underline{A}$ one can show that the converse is also true.

Let α be a morphism in $\underline{M}$. Denote by $d\alpha$ its domain and by $r\alpha$ its range.

(6) Theorem [11] 2.15. Let $\underline{M}$ and $\underline{A}$ be categories, $\underline{A}$ with direct sums ($\underline{M}$ need not be small). Then for every func-

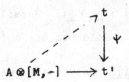

tor $t : \underline{M} \to \underline{A}$ there is a direct limit representation

$$t = \varinjlim td\alpha \otimes [r\alpha, -]$$

The objects of the corresponding index category are the morphisms of $\underline{M}$. The morphism set $[\alpha, \alpha']$ consists of one element if either $\alpha' = id_{d\alpha}$ or $\alpha' = id_{r\alpha}$. Otherwise it is empty (cf. [12] 2.19).

__Proof:__ We give an outline, making use of the adjointness property of $\otimes$, the dual of [12] 2.21 and the usual Yoneda lemma. For every $X \in \underline{M}$ and $Y \in \underline{A}$ there are natural isomorphisms

$$\left[\left\{\varinjlim td\alpha \otimes [r\alpha, X]\right\}, Y\right] \cong \varprojlim \left[td\alpha \otimes [r\alpha, X], Y\right] \cong$$

$$\varprojlim \left[[r\alpha, X], [td\alpha, Y]\right] \cong \left[[-, X], [t-, Y]\right] \cong [tX, Y]$$

This shows that $tX = \varinjlim td\alpha \otimes [r\alpha, X]$.

Q.E.D.

(7) __Remark.__[5] Moreover, one can show that the Yoneda functor

$$Y : \underline{M}^{opp} x \underline{A} \longrightarrow [\underline{M}, \underline{A}], \quad M \times A \rightsquigarrow A \otimes [M, -],$$

shares with the usual Yoneda embedding the property that it is

---

5) The properties (2), (5) and (7) show that generalized representable functors are a reasonable substitute for hom-functors in an arbitrary functor category $[\underline{M}, \underline{A}]$. In the following we omit "generalized" and call a functor $A \otimes [M, -] : \underline{M} \to \underline{A}$ representable. In order not to confuse Eilenberg-MacLane's notion of a "representable" functor with ours, we use quotation marks for the former. We will show later that the two notions are closely related.

dense (cf. Ulmer [12]) or adequate in the sense of Isbell. In otherwords, every functor $t \in [\underline{M},\underline{A}]$ is the direct limit of representables over the canonical index category which is the comma category (Y,t) .

(8) <u>Definition (Kan extension [8]).</u> Let $J : \underline{M} \to \underline{C}$ and $t : \underline{M} \to \underline{A}$ be functors. The right Kan J-extension of t is a functor $E_J(t) : \underline{C} \to \underline{A}$ such that for any functor $s : \underline{C} \to \underline{A}$ there is a bijection

$$[E_J(t),s] \cong [t,s \cdot J] ,$$

natural in s .

Dually, the left Kan J-extension of t is a functor $E^J(t) : \underline{C} \to \underline{A}$ such that there are natural bijections

$$[s,E^J(t)] \cong [s \cdot J,t]$$

It follows immediately that $E_J(t)$ and $E^J(t)$ are determined up to equivalence. We are mainly concerned with $E_J(t)$ and in short we call it the Kan extension. We leave it to the reader to state the dual theorems for $E^J(t)$. If J is full and faithful, it follows from (10) below that $E_J(t) \cdot J \cong t$, i.e. $E_J(t)$ is an extension of t . This explains the terminology.

(9) <u>Lemma.</u> Let $J : \underline{M} \to \underline{C}$ be a functor. The Kan extension of a representable functor $A \otimes [M,-] : \underline{M} \to \underline{A}$ is

$$A \otimes [JM,-] : \underline{C} \to \underline{A}$$

Likewise for an infinite sum $\bigoplus_i (A_i \otimes [M_i,-])$ the Kan extension is

$$\bigoplus_i (A_i \otimes [JM_i,-]) : \underline{C} \to \underline{A}$$

Similarly, the Kan extension of a hom-functor $[M,-] : \underline{M} \to \underline{S}$ is $[JM,-] : \underline{C} \to \underline{S}$.

Proof. For every functor $s : \underline{C} \to \underline{A}$ the Yoneda lemma (2) gives rise to bijections

$$\left[A \otimes [JM,-],s\right] \cong [A,sJM] \cong \left[A \otimes [M,-],s \cdot J\right]$$

which are natural in s . The second half can be established similarly.

Q.E.D.

Assume that $E_J(t)$ exists for every functor $t : \underline{M} \to \underline{A}$. Then the Kan extension $E_J : [\underline{M},\underline{A}] \to [\underline{C},\underline{A}]$ is left adjoint to the restriction $[\underline{C},\underline{A}] \to [\underline{M},\underline{A}]$, $s \rightsquigarrow s \cdot J$. Thus E_J preserves direct limits. This is also true if E_J is not defined everywhere.

(10) Theorem. Let $J : \underline{M} \to \underline{C}$ and $t : \underline{M} \to \underline{A}$ be functors. Assume $\underline{A}$ has arbitrary sums. Then $E_J(t) : \underline{C} \to \underline{A}$ exists iff $\varinjlim td\alpha \otimes [Jr\alpha,-]$ exists and

$$E_J(t) = \varinjlim td\alpha \otimes [Jr\alpha,-]$$

is valid.

(11) Corollary (Kan [8]). If $\underline{M}$ is small and $\underline{A}$ has direct limits, then for every functor $t : \underline{M} \to \underline{A}$ the Kan extension exists.

Proof of (10). By (6) there is a representation $t = \varinjlim td\alpha \otimes [r\alpha,-]$. Assume that $\varinjlim td\alpha \otimes [Jr\alpha,-]$ exists. From (9) it follows for every functor $s : \underline{C} \to \underline{A}$ that

$$\left[\varinjlim td\alpha \otimes [Jr\alpha,-],s\right] \cong \varprojlim \left[td\alpha \otimes [Jr\alpha,-],s\right] \cong$$

$$\cong \varprojlim \left[td\alpha \otimes [r\alpha,-],s \cdot J\right] \cong \left[\varinjlim td\alpha \otimes [r\alpha,-],s \cdot J\right] \cong [t,s \cdot J]$$

Hence $E_J(t)$ exists and $E_J(t) \cong \varinjlim td\alpha \otimes [Jr\alpha,-]$. Conversely, if $E_J(t)$ exists, then by reading the above isomorphism in the opposite direction, it follows that $E_J(t)$ is the limit $\varinjlim td\alpha \otimes [Jr\alpha,-]$. [6)]

(12) Let $J : \underline{M} \to \underline{C}$ be the full inclusion of a full small subcategory (referred to as models) and let $\underline{A}$ be an abelian category with sums and enough projectives. As shown in (5), a representable functor $P \otimes [M,-] : \underline{M} \to \underline{A}$ is projective iff P is projective in $\underline{A}$. Choose for every functor $t : \underline{M} \to \underline{A}$ and

---

6) The theorem shows that the existence of Kan extensions is equivalent with the existence of certain direct limits in $\underline{A}$ whose index systems are not sets but proper classes. From this it is fairly obvious that not every functor admits a Kan extension. For a counterexample see [13] p.6. Sometimes special properties of $\underline{M}$ or t guarantee the existence of $E_J(t)$. For instance, if t is a small direct limit of representable functors, then one can show, as in (10), that $E_J(t)$ exists, provided the range of t has direct limits.

every $M \in \underline{M}$ an epimorphism $P_M \to tM$, where P_M is projective.[7] By means of the Yoneda lemma $\left[P_M \otimes [M,-],t\right] \cong [P_M,tM]$ the family of epimorphisms determines a natural transformation $\psi(t) : \bigoplus_{M \in \underline{M}} \left(P_M \otimes [M,-]\right) \longrightarrow t$ which can be easily shown to be epimorphic (use the Yoneda lemma (2)). Thus t is projective in $[\underline{M},\underline{A}]$ iff it is a direct summand of a sum of projective representables. If t is the restriction of a functor $T : \underline{C} \to \underline{A}$, then this family and the Yoneda isomorphisms $\left[P_M \otimes [JM,-],T\right] \cong [P_M,TJM]$ determine also a natural transformation $\varphi(t) : \bigoplus_{M \in \underline{M}} \left(P_M \otimes [JM,-]\right) \longrightarrow T$.

(13) Theorem. Let $T : \underline{C} \to \underline{A}$ be a functor. The following are equivalent:

i) T is "representable" in the sense of Eilenberg-MacLane. (We use quotation marks to distinguish their notion of representability from ours.)

ii) $T \cdot J$ is projective in $[\underline{M},\underline{A}]$ and the Kan extension of $T \cdot J$ is T, i.e. $E_J(T \cdot J) = T$.

Moreover, the restriction $R_J : [\underline{C},\underline{A}] \to [\underline{M},\underline{A}]$ sets up an isomorphism between the "representable" functors $[\underline{C},\underline{A}]$ and the projective functors in $[\underline{M},\underline{A}]$. The inverse is given by the Kan extension $E_J : [\underline{M},\underline{A}] \longrightarrow [\underline{C},\underline{A}]$. (This shows that the notion of a "representable" functor (Eilenberg-MacLane) and a

---

7) If $\underline{A}$ is the category of abelian groups, one can choose P_M to be the free abelian group on tM. It is instructive for the following to have this example in mind. It links our approach with the one of Eilenberg-MacLane [6].

representable functor (Ulmer [11]) are closely related. Actually the functor T is also projective in $[\underline{C},\underline{A}]$, but this is irrelevant in the following.)

Proof. Eilenberg-MacLane call a functor $T : \underline{C} \to \underline{A}$ "representable" if the above natural transformation $\varphi(t) : \bigoplus_M (P_M \otimes [JM,-]) \to T$ admits a section σ (in other words, T is a direct summand of $\bigoplus_M (P_M \otimes [JM,-])$. Since $E_J(P_M \otimes [M,-]) = P_M \otimes [JM,-]$ (cf. (9)) and J is a full inclusion, the composite $E_J \cdot R_J : [\underline{C},\underline{A}] \to [\underline{M},\underline{A}] \to [\underline{C},\underline{A}]$ maps the sum $\bigoplus_M (P_M \otimes [JM,-])$ on itself. Obviously the same holds for a direct summand T of $\bigoplus_M (P_M \otimes [JM,-])$, i.e. $E_J(T \cdot J) = T$ is valid. By (12) the restriction $T \cdot J$ is projective. Conversely, if $T \cdot J$ is projective and $E_J(T \cdot J) = T$ holds, then the epimorphism $\psi(t) : \bigoplus_M (P_M \otimes [M,-]) \to T \cdot J$ splits and so does its Kan extension $\varphi(t) : \bigoplus_M (P_M \otimes [JM,-]) \to T$. It follows easily from this and (12) that the restriction functor R_J and the Kan extension E_J set up an isomorphism between the "representable" functors $\underline{C} \to \underline{A}$ and the projective functors $\underline{M} \to \underline{A}$.

Q.E.D.

From this we obtain immediately the following:

(14) Theorem. Let $T_* : \underline{C} \to \underline{A}$ be a positive complex [8] of functors together with an augmentation $T_* \to T_{-1}$. The following are equivalent:

8) This means that $T_n = 0$ for negative n. In the following we abbreviate "positive complex" to "complex".

i) The functors T_n are "representable" for $n \geq o$ and the augmented complex $T_* \cdot J \to T_{-1} \cdot J \to 0$ is exact.

ii) $T_* \cdot J$ is a projective resolution of $T_{-1} \cdot J$ and $E_J(T_n \cdot J) = T_n$ holds for $n \geq o$.

(15) __Corollary.__ If $\bar{T}_* \to \bar{T}_{-1}$ is another augmented complex satisfying i) such that $T_{-1} \cdot J \cong \bar{T}_{-1} \cdot J$ holds, then T_* and $\bar{T}_*$ are homotopically equivalent. Every chain map $T_* \to \bar{T}_*$ is a homotopy equivalence, provided its restriction on $\underline{M}$ is compatible with the augmentation isomorphism $T_{-1} \cdot J \cong \bar{T}_{-1} \cdot J$. Moreover, the n-th homology of T_* (and $\bar{T}_*$) is the value of the n-th left derived functor of E_J at t, where $T_{-1} \cdot J \cong t \cong \bar{T}_{-1} \cdot J$ (i.e. $H_n(T_*) \cong L_n E_J(t)$).

__Proof of (15).__ By (14)ii) the complexes $T_* \cdot J$ and $\bar{T}_* \cdot J$ are projective resolutions of t, and hence there is a homotopy equivalence $T_* \cdot J \simeq \bar{T}_* \cdot J$. Applying the Kan extension yields $T_* \simeq \bar{T}_*$. Clearly the restriction of every chain map $f_* : T_* \to \bar{T}_*$ on $\underline{M}$ is a homotopy equivalence $f_* J : T_* J \simeq \bar{T}_* J$ provided $f_* J$ is compatible with the augmentation isomorphism $T_{-1} \cdot J \cong \bar{T}_{-1} \cdot J$. Applying the Kan extension E_J on $f_* J$ yields again f_*. Hence f_* is also a homotopy equivalence. By standard homological algebra $L_n E_J(t) \cong H_n E_J(T_* \cdot J) = H_n T_*$ holds.

(16) __Remark.__ The method of acyclic models can be dualized by replacing the representable functors $A \otimes [M,-]$ by corepresent-

able functors $\underline{[-,A]}\cdot[-,M]$ and projective functors $P\otimes[M,-]$ by injective functors $\underline{[-,I]}\cdot[-,M]$ ($I\in\underline{A}$ injective, cf. [11] intro.). Thereby the right Kan extension E_J has to be replaced by the left Kan extension E^J (cf. (8)). Assume $\underline{A}$ has enough injectives. A functor $T:\underline{C}\to\underline{A}$ is said to be "corepresentable" (in the sense of Eilenberg-MacLane) iff the morphism $\varphi: T\to\prod_M\underline{[-,I_M]}\cdot[-,M]$ has a left inverse σ, i.e. $\sigma\cdot\varphi=\mathrm{id}$. (As in (12) φ is associated with a family of injections $\{TM\to I_M\}_{M\in\underline{M}}$, where $I_M\in\underline{A}$ is injective.) With this, one can easily dualize (13)-(15) and also the considerations below. Of course the notion of acyclicity is self-dual.

(17) Remark. Roughly speaking, the above shows that the method of acyclic models is the standard procedure in homological algebra to compute the left derived functors of the Kan extension by means of projectives. It is well known that the left derived functors of E_J can be computed not only with projectives but more generally by E_J-acyclic resolutions.[9] This leads to a useful generalization of acyclic models. For the considerations below, one can drop the assumption that $\underline{A}$ has sums and enough projectives and assume instead that the Kan extension and its left derived functors exist (for instance, if $\underline{A}$ is a Grothendieck AB4) category (cf. [7]) or if $\underline{M}$ consists of the projectives of a cotriple, cf. [3]).

---

9) Recall that $t\in[\underline{M},\underline{A}]$ is called E_J-acyclic if $L_nE_J(t)=0$ for $n>0$.

(18) Definition. A functor $T : \underline{C} \to \underline{A}$ is called "weakly representable" iff $E_J(T\cdot J) = T$ and $L_n E_J(T\cdot J) = 0$ for $n > 0$.[10] Clearly a "representable" functor is "weakly representable". If $\underline{A}$ is AB4), then every direct summand of a sum $\bigoplus_i (A_i \otimes [JM_i,-])$ is "weakly representable", where $M_i \in \underline{M}$ and $A_i \in \underline{A}$. Further examples are Λ-representable functors in the sense of M. André [1] p.11.

(19) Theorem. Let $T_*, \bar{T}_* : \underline{C} \to \underline{A}$ be complexes of "weakly representable" functors together with augmentations $T_* \to T_{-1}$ and $\bar{T}_* \to \bar{T}_{-1}$ such that $T_{-1}\cdot J \cong \bar{T}_{-1}\cdot J$ is valid and the augmented complexes $T_*\cdot J \to T_{-1}\cdot J \to 0$ and $\bar{T}_*\cdot J \to \bar{T}_{-1}\cdot J \to 0$ are exact. Then $H_n(T_*) \cong L_n E_J(t) \cong H_n(\bar{T}_*)$ holds for $n \geqslant 0$, where t is a functor isomorphic to $T_{-1}\cdot J$ or $\bar{T}_{-1}\cdot J$. Every chain map on the models $T_*\cdot J \to \bar{T}_*\cdot J$ can be uniquely extended to a chain $T_* \to \bar{T}_*$ by the Kan extension $E_J : [\underline{M},\underline{A}] \to [\underline{C},\underline{A}]$. In other words, the restriction of chain maps $[T_*,\bar{T}_*] \to [T_*\cdot J, \bar{T}_*\cdot J]$ is a bijection. Moreover, every chain map $f_* : T_* \to \bar{T}_*$ induces a homology isomorphism, provided its restriction on $\underline{M}$ is compatible with the augmenta-

---

10) There is a useful modification of the notion "weakly representable". Let $\mathcal{P}$ be a proper class of short exact sequences in $[\underline{M},\underline{A}]$ in the sense of MacLane [9] p.367, for instance, the class of objectwise split exact sequences. Then the condition $L_n E_J(T\cdot J) = 0$ can be replaced by $\mathcal{P}\text{-}L_n E_J(T\cdot J) = 0$, where $\mathcal{P}\text{-}L_*$ denotes the left derived functors with respect to $\mathcal{P}$. If the complexes $\bar{T}_*$ and T_* in (19) below have the property that $\bar{T}_* J \to \bar{T}_{-1} J \to 0$ and $T_* J \to T_{-1} J \to 0$ are $\mathcal{P}$-exact, then the statements (19)-(22) also hold if $L_* E_J$ is replaced by $\mathcal{P}\text{-}L_* E_J$.

tion $T_{-1}\cdot J \cong \bar{T}_{-1}\cdot J$.

Proof. The first half is standard homological algebra (e.g. see Grothendieck [7], Röhrl [10]). The restriction map $[T_*,\bar{T}_*] \to [T_*\cdot J,\bar{T}_*\cdot J]$ is a bijection because it is the adjunction isomorphism $[T_*,\bar{T}_*] = [E_J(T_*\cdot J),\bar{T}_*] \cong [T_*\cdot J ,\bar{T}_*\cdot J]$ (cf. (8)). Hence for every chain map $f_* : T_* \to \bar{T}_*$ the equation $E_J(f_*J) = f_*$ holds, where $f_*J : T_*\cdot J \to \bar{T}_*\cdot J$ is the restriction of f_* on $\underline{M}$. The mapping cône C_{f_*} of f_* is the Kan extension of the mapping cône of f_*J [11]. The latter consists of E_J-acyclic objects and is exact because by assumption $f_*J : T_*\cdot J \to \bar{T}_*\cdot J$ is a homology isomorphism. Hence C_{f_*} is also exact and $f_* : T_* \to \bar{T}_*$ induces a homology isomorphism.

(20) Remark. It is clear from the above that T_* and $\bar{T}_*$ are homotopically equivalent iff $T_*\cdot J$ and $\bar{T}_*\cdot J$ are. In general there need not be a homotopy equivalence between T_* and $\bar{T}_*$. In practice this lack can be compensated by the following:

Let $F : \underline{A} \to \underline{A}'$ be an additive functor, $\underline{A}'$ abelian, and assume that the objects T_nC and $\bar{T}_nC$ are F-acyclic for $n \geq 0$ and $C \in \underline{C}$. Then $Ff_* : FT_* \to F\bar{T}_*$ still induces a homology isomorphism, where $f_* : T_* \to \bar{T}_*$ is as in (19). This can be

11) For the definition of the mapping cône and its elementary properties, we refer to Dold [4]. Recall that a chain map g_* between complexes induces a homology isomorphism iff the mapping cône of g_* is an exact complex.

proved in the same way as above by means of the mapping cone technique[12]. If either $\underline{A}$ is AB4) and has enough F-acyclic objects [13] or $\underline{A}$ has sums and enough projectives, then one can construct a resolution $P_*(t)$ of t consisting in each dimension of a sum of representable functors which can be embedded into a diagram

$$\begin{array}{ccccc} T_*J & \longleftarrow & P_*(t) & \longrightarrow & \bar{T}_*\cdot J \\ \downarrow & & \downarrow & & \downarrow \\ T_{-1}\cdot J & \cong & t & \cong & \bar{T}_{-1}\cdot J \\ \downarrow & & \downarrow & & \downarrow \\ 0 & & 0 & & 0 \end{array}$$

One can choose $P_n(t) = \bigoplus_M (Q_M \otimes [M,-])$ where Q_M is either F-acyclic (first case) or projective (second case). Moreover the values of $E_JR_n(t) : \underline{C} \to \underline{A}$ are F-acyclic for $n \geq o$. Clearly the complex $E_JP_*(t) \to E_J(t)$ is "weakly representable" and acyclic on the models. Thus we obtain from the preceding argument homology isomorphisms $H_n(FT_*) \cong H_n(FE_JP_*(t)) \cong$ $\cong H_n(F\bar{T}_*)$.

The theorem below, which we state without proof, gives further information about this problem.

(21) Theorem. Let $T : \underline{C} \to \underline{A}$ be a "weakly representable"

---

12) In the examples, the objects T_nC and $\bar{T}_nC$ are usually projectives.

13) i.e. every object in $\underline{A}$ is a quotient of an object Q such that $L_nF(Q) = 0$ for $n > o$.

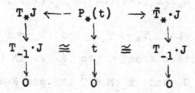

functor and let $F : \underline{A} \to \underline{A}'$ be an additive functor with a right adjoint. Assume either that $\underline{A}$ and $\underline{A}'$ are AB4) categories with enough F-acyclic objects or that they have sums and enough projectives.
Then $F\cdot T : \underline{C} \to \underline{A}'$ is "weakly representable" iff the objects TC are F-acyclic for $C \in \underline{C}$. (This follows from the equations $L_*E_J(F\cdot T\cdot J)(C) \cong L_*F\{E_J(T\cdot J)(C)\} = L_*F(TC)$, cf. [14] (19).)

(22) <u>Corollary.</u> Let $T_* \to T_{-1}$ and $\bar{T}_* \to \bar{T}_{-1}$ be augmented complexes of functors from $\underline{C}$ to $\underline{A}$ as in (19) such that the objects T_nC, $\bar{T}_nC$ and $T_{-1}M \cong \bar{T}_{-1}M$ are F-acyclic for $n \geq o$ and $C \in \underline{C}$ and $M \in \underline{M}$. Then it follows from (19) that $H_n(FT_*) \cong L_nE_J^{!}(F\cdot t) \cong H_n(F\bar{T}_*)$, where $T_{-1}\cdot J \cong t \cong \bar{T}_{-1}\cdot J$ and $E_J^{!} : [\underline{M},\underline{A}'] \to [\underline{C},\underline{A}']$ denotes the Kan extension. Clearly the other assertions in (19) also hold for $FT_* \to FT_{-1}$ and $F\bar{T}_* \to F\bar{T}_{-1}$.

(23) <u>Remark.</u> The isomorphism $H_n(T_*) \cong H_n(\bar{T}_*)$ in (19) can also be obtained from André's computational device for his homology functors $H_*(\, ,-) : [\underline{M},\underline{A}] \to [\underline{C},\underline{A}]$, provided $\underline{A}$ is AB4) (cf. [1] p.7). One can show that the functors $H_*(\, ,-) : [\underline{M},\underline{A}] \to [\underline{C},\underline{A}]$ coincide with $L_*E_J(-) : [\underline{M},\underline{A}] \to [\underline{C},\underline{A}]$, in particular $H_o(\, ,-) = E_J(-)$ (cf. [14], (4) this volume). Thus an augmented complex of functors $T_* \to T_{-1}$ satisfies André's condition in [1] prop. 1.5 iff it is acyclic on the models and T_n is "weakly representable" for $n \geq o$. Hence prop. 1.5 in [1] implies that $H_n(T_*) \cong H_n(-,T_{-1}\cdot J) \cong H_n(\bar{T}_*)$.

This shows that André's computational method is actually a generalization of acyclic models, the notion "representable" being replaced by "weakly representable". Barr-Beck [3] §XI used this method to improve their original version of acyclic models in [4], but all considerations were carried out in the functor category $[\underline{C},\underline{A}]$ without using the Kan extension. Their presentation in [3] §XI made me realize the relationship between acyclic models and Kan extensions. Before we discuss their approach, we sketch how the technique of acyclic models works when $\underline{M}$ is not small.

(24) Let $\underline{M}$ be a full but not necessarily small subcategory of $\underline{C}$ and let $\underline{A}$ be an abelian category with sums and enough projectives. Denote by $\widetilde{\underline{Ab}}.\underline{Gr}.$ the category of abelian groups, the underlying sets of which belong to a sufficiently large universe V which contains U (cf. intro.) and the sets of objects $\underline{M}$, $\underline{C}$ and $\underline{A}$. Let $\tilde{\underline{A}}$ be the category $\mathrm{Cont}_U(\underline{A}^{opp},\widetilde{\underline{Ab}}.\underline{Gr}.)$ of contravariant functors from $\underline{A}$ to $\widetilde{\underline{Ab}}.\underline{Gr}.$ which take U-small direct limits in inverse limits. The Yoneda embedding $I : \underline{A} \to \tilde{\underline{A}}$ is exact and preserves U-direct limits. Note that $\tilde{\underline{A}}$ has V-direct limits and enough projectives. The projectives are V-sums of hom-functors $[-,P]$, where $P \in \underline{A}$ is projective. Using the V-completion $\tilde{\underline{A}}$, one can proceed as before, and (12)-(22) carry over to this case. Define a functor $T : \underline{C} \to \underline{A}$ to be "representable" if $\varphi(t) : \bigoplus_M (P_M \otimes [JM,-]) \to I\cdot T$ has a section (cf. (12)). A

functor $T : \underline{C} \to \underline{A}$ is "representable" iff the composite $\underline{M} \xrightarrow{J} \underline{C} \xrightarrow{T} \underline{A} \xrightarrow{I} \breve{\underline{A}}$ is projective in $[\underline{M},\widetilde{\underline{A}}]$ and $\widetilde{E}_J(I\cdot T\cdot J) = I\cdot T$ holds, where $\widetilde{E}_J : [\underline{M},\widetilde{\underline{A}}] \to [\underline{C},\widetilde{\underline{A}}]$ is the Kan extension. Two complexes of "representable" functors $T_*,\bar{T}_* : \underline{C} \to \underline{A}$ with augmentations $T_* \to T_{-1}$ and $\bar{T}_* \to \bar{T}_{-1}$ are homotopically equivalent, provided the augmented complexes $T_*\cdot J \to T_{-1}\cdot J \to 0$ and $\bar{T}_*\cdot J \to \bar{T}_{-1}\cdot J \to 0$ are exact and $T_{-1}\cdot J \cong \bar{T}_{-1}\cdot J$ is valid. Every chain map $T_* \to \bar{T}_*$ is a homotopy equivalence, provided its restriction on $\underline{M}$ is compatible with the augmentation isomorphism $T_{-1}\cdot J \cong \bar{T}_{-1}\cdot J$. Moreover $H_n(T_*)$ is isomorphic with the value of $L_n\widetilde{E}_J : [\underline{M},\widetilde{\underline{A}}] \to [\underline{C},\widetilde{\underline{A}}]$ at $I\cdot t$, where $T_{-1}\cdot J \cong t \cong \bar{T}_{-1}\cdot J$. (Note that $[\underline{M},\widetilde{\underline{A}}]$ has enough projectives.) As before, the notion of a "representable" functor $T : \underline{C} \to \underline{A}$ can be replaced by the notion of a "weakly representable" functor. The latter can mean either $E_J(T\cdot J) = T$ and $L_nE_J(T\cdot J) = 0$ or $\widetilde{E}_J(I\cdot T\cdot J) = I\cdot T$ and $L_n\widetilde{E}_J(I\cdot T\cdot J) = 0$ for $n > 0$. Then T_* and $\bar{T}_*$ still have isomorphic homology and $H_n(T_*) \cong L_nE_J(t)$ or $I\cdot H_n(T_*) \cong H_n(I\cdot T_*) \cong L_n\widetilde{E}_J(I\cdot t)$ hold. Likewise every chain map $f_* : T_* \to \bar{T}_*$ induces a homology isomorphism, provided its restriction on $\underline{M}$ is compatible with the augmentation isomorphism $T_{-1}\cdot J \cong \bar{T}_{-1}\cdot J$. If $F : \underline{A} \to \underline{A}'$ is an additive functor, then $Ff_* : FT_* \to F\bar{T}_*$ is also a homology isomorphism, provided T_nC and $\bar{T}_nC$ are F-acyclic for $C \in \underline{C}$ and $n \geq 0$, etc.

(25) Barr-Beck [2] gave a version of acyclic models in which

the model category $\underline{M}$ is not small. For the sake of completeness, we briefly recall their approach. Let $\mathbb{G}$ be a cotriple in category $\underline{C}$ (cf. [3]) and $\underline{A}$ be an abelian category. Denote by $J : \underline{M} \to \underline{C}$ the inclusion of the full subcategory consisting of the objects GC, where $C \in \underline{C}$. A complex of functors $T_* : \underline{C} \to \underline{A}$ with an augmentation $T_* \to T_{-1}$ is called acyclic on the models $\underline{M}$ iff $T_* \cdot G \to T_{-1} \cdot G$ has a contraction. The functor T_n is called representable iff the canonical natural transformation $T_n G \xrightarrow{T\varepsilon} T_n$ admits a section $\sigma : T_n \to T_n G$, where $\varepsilon : G \to id_{\underline{C}}$ is the co-unit of the cotriple. Barr-Beck proved in [2] that two complexes of representable functors T_*, $\bar{T}_* : \underline{C} \to \underline{A}$ with isomorphic augmentations $T_{-1} \cong \bar{T}_{-1}$ are homotopically equivalent if $T_* \to T_{-1}$ and $\bar{T}_* \to \bar{T}_{-1}$ are acyclic on the models.[14] In particular T_* and $\bar{T}_*$ have isomorphic homology. Later they pointed out in [3] §XI that for the latter (i.e. $H_n(T_*) \cong H_n(\bar{T}_*)$) their notions of representability and acyclicity can be considerably weakened. Denote by $H_*(-,T)_{\mathbb{G}} : \underline{C} \to \underline{A}$ the cotriple homology associated with $\mathbb{G}$ and a coefficient functor $T : \underline{C} \to \underline{A}$. Their modified definition of representability is: $H_j(-,T_n)_{\mathbb{G}} = 0$ for $j > 0$ and $H_0(-,T_n)_{\mathbb{G}} = T_n$; of acyclicity: $T_*M \to T_{-1}M \to 0$

---

14) It should be noted that this homotopy equivalence cannot be obtained by the method outlined in (24) if one takes $J : \underline{M} \to \underline{C}$ to be the full inclusion of the objects of the form GC, where $C \in \underline{C}$. To obtain the homotopy equivalence one has to define $J : \underline{M} \to \underline{C}$ in a more elaborate way which we omit in view of the more general situation we discuss below.

is an exact complex for $M \in \underline{M}$. These definitions are obviously weaker because $H_o(-,T\cdot G)_G = T\cdot G$ and $H_j(-,T\cdot G)_G = 0$ hold[15] for every functor $T : \underline{C} \to \underline{A}$ and $j > o$. To make the connection between this (generalized) version of acyclic models and classical homological algebra we first recall that $H_o(\ ,-)_G : [\underline{C},\underline{A}] \to [\underline{C},\underline{A}]$ is the composite of the restriction $R_J : [\underline{C},\underline{A}] \to [\underline{M},\underline{A}]$ with the Kan extension $E_J : [\underline{M},\underline{A}] \to [\underline{C},\underline{A}]$ (cf. [14] this volume). Moreover $H_*(\ ,-)_G : [\underline{C},\underline{A}] \to [\underline{C},\underline{A}]$ is the composite of R_J with $L_*E_J : [\underline{M},\underline{A}] \to [\underline{C},\underline{A}]$. This shows that the modified notions of acyclicity and representability of Barr-Beck [3] §XI coincide with "acyclic" and "weakly representable" as defined in (18). Hence their method of acyclic models is essentially the standard procedure in homological algebra to compute the left derived functor of the Kan extension $E_J : [\underline{M},\underline{A}] \to [\underline{C},\underline{A}]$ by means of E_J-acyclic resolutions. We leave it to the reader to state theorems analogous to (19)-(22).

---

15) Clearly the same is valid for every direct summand of $T\cdot G$.

BIBLIOGRAPHY

[1] André, M., *Méthode simpliciale en algèbre homologique et algèbre commutative*, (Lecture notes in Mathematics, #32), Springer, 1967.

[2] Barr, M. and J. Beck, "Acyclic models and triples", in; *Conference on Categorical Algebra*, pp.336-343, Springer, (1966).

[3] ________. "Homology and standard constructions", (to appear in Lecture notes in Mathematics.)

[4] Dold, A., "Zur Homotopietheorie der Kettenkomplexe", *Math. Annalen*, 140; 278-298, (1960).

[5] ________, S. MacLane, U. Oberst, "Projective classes and acyclic models", in; A. Dold, Heidelberg and Eckmann (eds.), *Reports of the Midwest Category Seminar*, (Lecture notes in Mathematics, #47); 78-91, Springer, 1967.

[6] Eilenberg, S. and S. MacLane, "Acyclic models", *Am. J. Math.*, 75; 189-199, (1953).

[7] Grothendieck, A., "Sur quelques points d'algèbre homologique", *Tohoku, Math. J.*, 9; 119-221, (1957).

[8] Kan, D., "Adjoint Functors", *Trans. Amer. Math. Soc.*, 87; 295-329, (1958).

[9] MacLane, S., *Homology*, Springer, 1963.

[10] Röhrl, H., "Satelliten halbexakter Funktoren", *Math. Zeitschrift*, 79; 193-223, (1962).

[11] Ulmer, F., "Representable functors with values in arbitrary categories", *J. of Algebra*, 8; 96-129, (1968).

[12] ________. "Properties of dense and relative adjoint functors", *J. of Algebra*, 8; 77-95, (1968).

[13] ________. "Properties of Kan extensions", *Mim. Notes*, E.T.H., (1966).

[14] ________. "Kan extensions, Cotriple and André (co)homology", in; Lecture notes in mathematics, this volume.

DERIVED CATEGORY AND POINCARÉ DUALITY

by

M. Zisman*

The notations for derived categories and derived functors are those of HARTSHORNE [3]. Capital letters X,Y, Z, design locally compact Haussdorff spaces; A is a ring with unity given once for all; we write ${}_AX$, ${}_AY$, for the categories of sheaves of left A-modules over X,Y. (If we do not suppose the ring A to be commutative, we generally must add the hypothesis that A is a Λ-flat Λ-algebra where Λ is a commutative ring.) In the following a sheaf always means a sheaf of A-modules.

1. THE FUNCTOR $f_!$

1.1

Let $f: X \longrightarrow Y$ be a continuous map, and let F be a sheaf over X. We define a sheaf $f_!F$ over Y in the following way: Given any open set U in Y, then $f_!F(U)$ is the set of all sections $s \in F(F^{-1}U)$ whose support $K \subset f^{-1}(U)$ is such that the restriction $f|K: K \longrightarrow U$ is a proper map.

*Institut Henri Poincaré, 11 Rue Pierre Curie, Paris.

It is easy to check that the presheaf

$$U \longrightarrow f_! F(U)$$

is in fact a sheaf.

1.2 Examples

1.2.1. If the space Y is a point, then we have $f_! F = \Gamma_c(X,F)$ where Γ_c stands for the set of all sections of F with compact support.

1.2.2. If $f: X \hookrightarrow Y$ is the inclusion of an open subset X into Y, then $f_! F$ is the extension F^Y of F to Y, putting $(F^Y)_y = 0$ for $y \notin X$.[1)]

1.2.3. Of course if $f: X \longrightarrow Y$ is a proper map, then $f_!$ is nothing but the usual direct image functor f_*.

1.3

Consider for a while the case where X is a differentiable manifold of dimension n, Y is a point, $f: X \longrightarrow Y$ the unique map from X to a point, and $A = \mathbb{R}$ the field of real numbers. It is well known that the deRHAM's resolution of the constant sheaf $\mathbb{R}$ over X

$$0 \longrightarrow \mathbb{R} \longrightarrow \Omega^\circ \longrightarrow \Omega^1 \longrightarrow \cdots \longrightarrow \Omega^n \longrightarrow 0$$

has the following property:

---

1) Given a sheaf F over a space X and a point $x \in X$, then we write F_x for the stalk of F over x.

Given any open set U in X we get

$$R^q f_!(\Omega^p{}_U) = H^q_c(U, \Omega^p) = 0$$

for any $q > 0$ and any $p \geq 0$.

More generally, if $f: X \longrightarrow Y$ is a continuous map and F a sheaf over X, we say that F is f-soft if the following relation holds:

$R^q f_! F_U = 0$ for any $q > 0$ and any U open in X.

In other words, F is f-soft if F_U is $f_!$-acyclic for any open set $U \subset X$. (Confer GODEMENT [2] for the notation F_U).

<u>1.4 The condition $\Delta(A)$</u>

We say that the condition $\Delta(A)$ holds for a map $f: X \longrightarrow Y$ iff

(i) f <u>is of finite cohomological dimension</u> (i.e. there exist $q_\circ > 0$ such that $R^q f_! F = 0$ for any sheaf F and any $q \geq q_\circ$).

(ii) There exist a finite resolution of the constant sheaf A over X

$$0 \longrightarrow A \longrightarrow \Omega^\circ \longrightarrow \Omega^1 \longrightarrow \dots \longrightarrow \Omega^r \longrightarrow 0$$

<u>where the sheaves Ω^p are f-soft and A-(right)-flat.</u>

The following three lemmas assert that there exist maps for which $\Delta(A)$ holds.

<u>1.4.1 Lemma.</u> If $f_!$ is of finite cohomological dimension and if f is notherian, then $\Delta(A)$ holds for f.

1.4.2 Lemma. Let A $\longrightarrow$ B be a ring homomorphism (which send the unity of A into the unity of B). If $\Delta(A)$ holds for f, then Δ (B) holds for f.

1.4.3 Lemma. Δ (A) is closed under change of basis and compositions. Moreover, if Δ (A) holds for $g \cdot f$, it holds for f.

Using these lemmas we easily see that $\Delta(A)$ holds for any map between (topological) manifolds and then, using any change of basis, for a lot of maps.

1.5 Theorem (Poincaré Duality)

Let X and Y be locally compact Haussdorff spaces and let f: X $\longrightarrow$ Y be a continuous map. Suppose that the condition $\Delta(A)$ holds for f. Then there exists a unique (up to natural isomorphism) functor $f^!: D^+(_AY) \longrightarrow D^+(_AX)$ provided with a bifunctorial isomorphism

$$(*) \quad R\,Hom\,(Rf_!F^\cdot, G^\cdot) \xrightarrow{\approx} R\,Hom\,(F^\cdot, f^!G^\cdot)$$

where $F^\cdot$ is any object in $D(_AX)$ and $G^\cdot$ any object in $D^+(_AY)$. (For a proof, see [4] exposé 4.) Since $H^oR\,Hom^\cdot$ is nothing but Hom_D, we get, taking the 0-cohomology:

$$Hom_{D(_AY)}\,(Rf_!F^\cdot{}_\circ G^\cdot) \approx Hom_{D(_AX)}\,(F^\cdot, f^!G^\cdot)$$

so that the functor $f^!$ is in some sense right adjoint to $Rf_!$.

1.5.1. The description of $f^!$ can be done in the following way.

Since the category $D^+({}_AY)$ is equivalent to the category whose objects are bounded below complexes of sheaves, each sheaf in the complex being injective (and whose maps are homotopy classes of maps between complexes), we may suppose that $G^\cdot$ is such a complex. Given a resolution $\Omega^\cdot$ as in 1.4 (ii), we put

$$(f^!G^\cdot)\ (U) = \mathrm{Hom}^\cdot(f_!\Omega^\cdot{}_U, G^\cdot)$$

The complex of presheaves

$$U \longrightarrow f^!G^\cdot(U)$$

is in fact a complex of sheaves, the complex we are looking for.

1.6

In order to make understandable the reason why theorem 1.5 is called a Poincaré duality theorem, we shall now compute both sides of our formula (*) in the case where X is a topological manifold of dimension n, Y is a point, and where we take $G^\cdot = A$.[2)]

Let $\mathcal{H}$ design the cohomology of a complex of sheaves. If x is a point in X, according to GODEMENT [2], 4.1, we have:

$$\mathcal{H}^q(f^!A)_x = \varinjlim H^q\ \mathrm{Hom}^\cdot(f_!\Omega^\cdot{}_U, I^\cdot A)$$

---

2) Recall that a complex of sheaves over a point is simply a complex of A-modules. Moreover, we write F for the complex which has 0 in any degree ≠ 0 and has F in degree 0.

where $I^{\bullet}A$ is an injective resolution of A, and where U runs through the filtered set of all open neighborhoods of x in X. Since X is a manifold, we are allowed to suppose $U = R^n$, and then we get

$$\mathcal{H}^q(f^! A)_x = \begin{cases} 0 \text{ if } q \neq -n \\ A \text{ if } q = -n \end{cases}$$

so that the complex of sheaves $f^! A$ is quasi-isomorphic to the complex which has 0 in any degree $\neq -n$ and has the usual orientation sheaf $\mathcal{O}$ of X in degree -n. Suppose now that the complex $F^{\bullet}$ is simply a sheaf F. The formula (*) gives an isomorphism

$$H^q \operatorname{Hom}^{\bullet}(R\Gamma_c(X,F), I^{\bullet}A) \approx \operatorname{Ext}^{q+n}(F,\mathcal{O})$$

i.e. a spectral sequence

$$E_2^{p,q} = \operatorname{Ext}^p(H_c^{-q}(X,F), A) \Rightarrow \operatorname{Ext}^{p+q+n}(F,\mathcal{O}) .$$

In particular, if A is a field, we have the usual isomorphism

$$\operatorname{Hom}(H_c^q(X,F), A) \approx \operatorname{Ext}^{n-q}(F,\mathcal{O}) .$$

2. PERFECT COMPLEXES OF SHEAVES

(Confer [4] exposé 9)

If we want to walk in the direction of a LEFSCHETZ formula for sheaves, we must introduce some finiteness conditions on the sheaves used in the theory;

more precisely we first define a perfect complex of A-modules as a complex which is bounded, free and of finite type in each degree and afterwards denote by $D^b_{f.f.\ell.}({}_AX)$ the full subcategory of $D^b({}_AX)$ whose objects are those complexes of sheaves which are quasi-isomorphic to bounded complexes whose sheafs are flat in each degree; (f.f.ℓ. stands for "finite flat length").

2.1 Theorem

Let $F^{\bullet}$ be a complex in $D^b_{f.f.\ell.}({}_AX)$ where X is a locally compact Haussdorff space such that the condition $\Delta(A)$ holds for the map $\pi_X: X \longrightarrow$ point. Then the following two conditions are equivalent:

For any point x in X,

(i) The local inverse system $U \longrightarrow R\Gamma_c(U,F^{\bullet})$ is essentially perfect.

(ii) The local direct system $U \longrightarrow R\Gamma(U,F^{\bullet})$ is essentially perfect.

(U runs through the set of all open neighborhoods of x in X.)

Recall that the conditions (i) and (ii) mean that, given $x \in U$, there exist two open neighborhoods V and V' of x, contained in U, such that the natural maps

$$R\Gamma_c(V,F^{\bullet}) \longrightarrow R\Gamma_c(U,F^{\bullet})$$

$$R\Gamma(U,F^{\bullet}) \longrightarrow R\Gamma(V',F^{\bullet})$$

factor (in the derived category D(A) of the category of

A-modules) through a perfect complex.

2.2 Definition

A complex of sheaves satisfying the conditions of the preceding theorem will be called a perfect complex (of sheaves).

Perfect complexes look like sheaves (or complexes of sheaves) satisfying the BOREL-WILDER equivalent conditions ([1] and [4]). In fact we have the

2.3 Theorem

Let X *be a space as in* 2.1 *and* $F^\bullet$ *be a perfect complex of sheaves over* X.

(i) *Given any compact subset* K *of* X *and any open subset* U *of* X *such that* $U \supset K$, *then the canonical morphism*

$$R\Gamma(U,F^\bullet) \longrightarrow R\Gamma(K,F^\bullet)$$

factors in D(A) *through a perfect complex.*

(ii) *Given two open subspaces* U *and* V *of* X *such that* $\overline{U} \subset V$ *and* $\overline{U}$ *compact, then the canonical morphism*

$$R\Gamma_c(U,F^\bullet) \longrightarrow R\Gamma_c(V,F^\bullet)$$

factors in D(A) *through a perfect complex.*

3. A FEW PROPERTIES OF $f^!$

(Confer [4] exposés 9, 10, 11)

3.1 The dual complex

Let X be a space as in 2.1, we define the dual complex $D_X(F^\cdot)$ of a complex of sheaves $F^\cdot$ by the formula

$$D_X(F^\cdot) = R\,\mathcal{H}om^\cdot(F^\cdot, \pi_X^! A)$$

where $\mathcal{H}om$ is the inner Hom in the category ${}_A X$ (see for example GODEMENT [2], 2.2). As in the usual cases, we get a natural morphism $F^\cdot \longrightarrow D_X D_X F^\cdot$ from $F^\cdot$ into its bidual.

3.1.1 Theorem (Biduality Theorem). Suppose the complex $F^\cdot$ is perfect. Then the complex $D_X F^\cdot$ is also perfect and the natural morphism $F^\cdot \longrightarrow D_X D_X F^\cdot$ is an isomorphism.

(In particular, if the constant sheaf A over X is perfect, so is $\pi_X^! A$: confer 1.6 for the case where X is a manifold.)

3.2

Consider a map $f: X \longrightarrow Y$ such that $\Delta(A)$ holds for f. In both cases

(i) $G^\cdot \in Ob\, D^b_{f.f.\ell.}({}_A Y)$, $F^\cdot \in Ob\, D^+({}_A Y_A)$

(ii) $G^\cdot \in Ob\, D^+({}_A Y)$, $F^\cdot \in Ob\, D^b_{f.f.\ell.}({}_A Y_A)$, $f^! F^\cdot \in D^b({}_A Y_A)$

we are able to construct a natural morphism

$$\phi: f^! F^{\cdot} \overset{L}{\otimes} f^* G^{\cdot} \longrightarrow f^! (F^{\cdot} \overset{L}{\otimes} G^{\cdot})$$

where $\overset{L}{\otimes}$ stands for the left derived functor of the tensor product and f^* is the usual inverse image functor. The next (and last) theorem gives a condition on f in order to involve that ϕ is an isomorphism.

3.2.1 Definition. A map $f: X \longrightarrow Y$ is locally a product if for each point $x \in X$ there exist an open neighborhood U in x, an open neighborhood V of f(x), and a topological space Z provided with an homeomorphism $U \longrightarrow V \times Z$ making the following diagram commutative

$$\begin{array}{ccc} U & \xrightarrow{\approx} & V \times Z \\ {\scriptstyle f|U}\downarrow & \swarrow {\scriptstyle pr_V} & \\ V & & \end{array}$$

3.2.2 Theorem. Under the following assumptions

a) $f: X \longrightarrow Y$ is locally a product

b) the condition $\Delta(A)$ holds for f

c) for any $y \in Y$, the constant sheaf A over $f^{-1}(y)$ is perfect, we have

(1) If $F^{\cdot}$ is an object in $D^b_{f.f.\ell.}({}_A Y)$ (resp. is perfect) then $f^! F^{\cdot}$ is an object in $D^b_{f.f.\ell.}({}_A X)$ (resp. is perfect)

(2) In both cases (i) and (ii) of 3.2 the morphism

$$\phi: f^! F^{\cdot} \overset{L}{\otimes} f^* G^{\cdot} \longrightarrow f^! (F^{\cdot} \overset{L}{\otimes} G^{\cdot})$$

is an isomorphism.

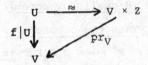

3.2.3 How to define ϕ (in case (ii); the construction in case (i) is quite different; however both constructions agree when $F^\cdot$ and $G^\cdot$ satisfy conditions (i) and (ii) together).

a) Because $f^! F^\cdot \overset{L}{\otimes} ?$ is left adjoint to $R\,\mathcal{H}om^\cdot(f^! F^\cdot, ?)$ since $f^! F^\cdot$ is bounded, it is enough to define a morphism

$$F^* G^\cdot \longrightarrow R\,\mathcal{H}om^\cdot(f^! F^\cdot, f^!(F^\cdot \overset{L}{\otimes} G^\cdot))$$

b) But f^* is left adjoint to Rf_* and then we just need a morphism

$$G^\cdot \longrightarrow Rf_* \, R\,\mathcal{H}om^\cdot(f^! F^\cdot, f^!(F^\cdot \overset{L}{\otimes} G^\cdot))$$

c) Now the Poincaré Duality theorem (using the functor $\mathcal{H}om^\cdot$ instead of $\mathrm{Hom}^\cdot$) gives an isomorphism

$$\begin{array}{c} Rf_* \, R\,\mathcal{H}om^\cdot(f^! F^\cdot, f^!(F^\cdot \overset{L}{\otimes} G^\cdot)) \\ \downarrow \approx \\ R\,\mathcal{H}om^\cdot(Rf_! f^! F^\cdot, F^\cdot \overset{L}{\otimes} G^\cdot) \end{array}$$

and so we seek a morphism

$$G^\cdot \longrightarrow R\,\mathcal{H}om^\cdot(Rf_! f^! F^\cdot, F^\cdot \overset{L}{\otimes} G^\cdot) ,$$

d) or, using once more the adjunction between $\overset{L}{\otimes}$ and $R\,\mathcal{H}om^\cdot$, a morphism

$$\phi' : Rf_! f^! F^\cdot \overset{L}{\otimes} G^\cdot \longrightarrow F^\cdot \overset{L}{\otimes} G^\cdot \quad .$$

e) at this stage we very happily remember the adjunction between $Rf_!$ and $f^!$: in fact this gives a natural morphism

$$\Phi : Rf_! f^! \longrightarrow \mathrm{Id}$$

so that we finally define ϕ' by the formula

$$\phi' = \phi(F^{\cdot}) \overset{L}{\otimes} Id(G^{\cdot}) .$$

REFERENCES

[1] BOREL, A. "Poincaré Duality in Generalized Manifolds", *Michigan Math. J.*, 4; 227-239, (1957).

[2] GODEMENT, *Theorie des faisceaux*, Hermann (Actualites scientifiques et industrielles, #1252), (1958).

[3] HARTSHORNE, *Residues and Duality*, (Lecture Notes in Mathematics, #20), Springer Verlag, (1966).

[4] VERDIER-ZISMAN, "Seminaire sur la formule de Lefschetz", (multigraphié) *Publication IRMA Dept de Mathematique Faculte des Sciences, Strasbourg*.

Offsetdruck: Julius Beltz, Weinheim/Bergstr.